W9-AGJ-252

SPACE, TIME, AND CREATION

MILTON K. MUNITZ

SPACE

TIME AND CREATION

Philosophical Aspects

of Scientific Cosmology

COLLIER BOOKS
NEW YORK, N.Y.

This Collier Books edition is published by
arrangement with The Free Press

First Collier Books edition: September, 1961

Collier Books are published by
the Crowell-Collier Publishing Company

Printed in the United States of America

Library of Congress Catalog Card No. 57-6747

That which is far off, and exceeding deep, who can find it out?
(Eccles. 7:24)

It is appropriate to approach the problems of cosmology with feelings of respect for their importance, of awe for their vastness, and of exultation for the temerity of the human mind in attempting to solve them. They must be treated, however, by the detailed, critical, and dispassionate methods of the scientist.

R. C. Tolman, *Relativity Thermodynamics and Cosmology*, 488

To My Wife

PREFACE

I WISH to thank the Ford Foundation (Fund for the Advancement of Education) for the award of a Faculty Fellowship for the academic year 1954-1955. It was during the leisure time thus provided that I was able to write the major portion of this book.

Chapter II, Section 5, draws upon some material previously published in an article entitled "One Universe or Many?" in the *Journal of the History of Ideas,* Volume XII (1951); Chapter V, Section 2 and Chapter VI, Section 2 include some material from an article "Scientific Method in Cosmology," which appeared in *Philosophy of Science,* Volume 19 (1952). Chapter IX is a slightly revised version of an article, "Creation and the New Cosmology" in the *British Journal for the Philosophy of Science,* Volume V (1954). Grateful acknowledgement is made to the editors of the above periodicals for permission to make use of the articles mentioned.

I am much indebted to Professor Sidney Hook for his many acts of generous support in my behalf and for his friendly encouragement. As Chairman of the Department of Philosophy of New York University he has afforded me the opportunity of originating and teaching a course in "Theories of the Universe" both in the Graduate School of Arts and Sciences and in Washington Square College. Much of the material here presented formed the basis for the subject matter of this course.

Professor Ernest Nagel of Columbia University, whose friendship I have treasured over a period of many years, has been another unfailing source of personal support and rewarding philosophic counsel. Few philosophers in my experience have shown as clearly as he has by his own work and teaching the valuable fruits to be had for practising philos-

7

ophy in the context of some concrete scientific subject matter.

I am also much indebted to Professor Albert Hofstadter of Columbia University for reading the manuscript in its entirety and for giving me the benefit of his acute criticisms. Again, I was particularly fortunate in being able to profit from the remarks of Professor G. C. McVittie of the University of Illinois, an outstanding authority in the field of scientific cosmology, who read much of the manuscript.

Three weeks before the death of Albert Einstein I had the very great privilege of a long conversation with him. It was both an unforgettable and an illuminating experience. A brief summary of his comments on that portion of my discussion which undertakes to criticize his views on space and on scientific knowledge generally, will be found in the appropriate section of the text below.

My colleague Professor Paul Edwards, as Philosophy Editor for The Free Press, has given much of his time and energy in arranging for the publication of this manuscript as well as for that of a companion volume entitled *Theories of the Universe: From Babylonian Myth to Modern Science.* I wish to thank him most heartily.

Finally, my wife, Dr. Lenore B. Munitz, has been of inestimable help both in giving me the benefit of her keen philosophic sense as well as in the day-to-day assistance of a more practical secretarial nature.

M. K. M.

Scarborough, New York
October 1956

Contents

PREFACE   7

I   INTRODUCTION—WHAT IS COSMOLOGY?   11

II   THE DEVELOPMENT OF COSMOLOGY
FROM MYTH TO SCIENCE   13
   1. Kinds and Stages of Progress in Cosmology   13
   2. Cosmogonical Myths: *The Enuma Elish*   16
   3. Proto-Scientific Cosmologies:
      Anaximander   21
   4. The Classical World-View: Aristotle   22
   5. The Universe of Stars in Modern Astron-
      omy   26
   6. Nebulae and Relativity   35

III   SOME QUESTIONS OF LOGIC IN COSMOLOGY   37
   1. Deduction of Extrapolation? Space and
      Time   38
   2. Importance of Philosophical Analysis   40

IV   THE NATURE OF SCIENTIFIC THEORY   41
   1. Functions of Theory   42
   2. Types of Theoretical Concepts   49
   3. Types of Theoretical Statements   52
   4. Critique of the Classical Philosophy of
      Science   54

V   THE OBSERVABLE UNIVERSE AND COSMOLOGICAL
MODELS   61
   1. Meaning of "the Universe"   61
   2. Relativistic Cosmologies   68

VI RATIONALISM 78
    1. The Argument from the Uniqueness of the
       Universe 80
    2. Milne's Kinematic Relativity 86

VII WORLD GEOMETRY 92
    1. Concepts of Physical Space 93
    2. The "Reality" of Space; Einstein's views 100
    3. The Metrical Structure of Space 104
    4. The Metric of the Universe as a Whole 112

VIII THE AGE OF THE UNIVERSE 117
    1. Time and the Use of Clocks in Physics 120
    2. "The Beginning" of the Universe 125

IX CONTINUOUS CREATION 134
    1. A New Cosmology: Bondi-Gold, Hoyle 134
    2. The "Physics" of Creation 135
    3. Creation and the Use of Scientific Method 137
    4. The Meaningfulness of "Creation" 140
    5. The Perfect Cosmological Principle 143

X SCEPTICISM 149
    1. The Challenge of Operationalism 149
    2. Kantian Dialectic 150
    3. Conclusion 154

INDEX 155

Chapter 1

Introduction—What Is Cosmology?

OUR SENSE of sight in disclosing to us the sweep of the
heavens and the patterns of change in its several parts, is
the most important stimulus to the perennial human search
for understanding the universe in which we live. "The sight"
Plato has Timaeus say at one point in his memorable discourse, "in my opinion is the source of the greatest benefit
to us, for had we never seen the stars, and the sun, and the
heaven, none of the words which we have spoken about the
universe would ever have been uttered. But now the sight of
day and night, and the months and the revolutions of the
years, have created number, and have given us a conception
of time, and the power of enquiring about the nature of the
universe; and from this source we have derived philosophy,
than which no greater good ever was or will be given by the
gods to mortal man." [1] But the sense of sight, as Plato is also
quick to note, whatever its benefits and glories, only feeds a
capacity for wonder and curiosity that it itself is unable to
satisfy. Men ask themselves questions which take them beyond what they immediately see. They strive for a complete
account, one that would fill out what they only fragmentarily
observe from their local stations in time and space. Above
all they seek for some pattern which will give intelligibility
and measure to what they perceive. As long as men are what
they are, equipped with senses and an inquiring intelligence,
this has been and will always be the case. The central interest
of cosmology, that of providing a satisfactory account of the
physical universe as a whole, is accordingly both an ancient
and a persistent one. Each culture and age has proposed its
own answers. Myth, philosophy, and science have each under-
taken to deal with the problem in some fashion or other.

An approach to cosmology from a logical or critical point
of view will seek to understand—both in the light of a study
of past efforts and particularly in terms of its most recent
scientific instances—what such efforts involve. Since cosmol-

[1] *Timaeus*, 47.

ogy undertakes to give us knowledge of the universe, the root question that critical philosophy asks is: *"In what does knowledge of the universe consist?"*—or, giving special emphasis, in turn, to the parts of this question, we may ask: "What does it mean to *know* the universe?" and, "To what do we refer when we speak of *the universe* as the object of our knowledge?" These questions are *not* to be satisfied by providing an answer in terms of some preferred cosmology. It is not the task of logic or critical philosophy to participate directly in the construction of a theory of the universe, or to undertake a technical evaluation of the variety of proposed accounts. Its task is rather that of an onlooker, at second remove, so to speak, from the subject matter with which cosmology deals. It seeks, not to understand the universe, but to understand in what such an understanding does, or might, consist. It would enlighten us about the *kind* of subject matter, problems, methods, and goals with which cosmology may be identified in its search for knowledge of the universe.

The general purpose of the following pages is to attempt to answer this broad critical question on the basis of an examination of representative cosmological schemes, both historical and contemporary, and the methods which lie behind their construction. Two themes will be selected for particular attention: (1) One of the special purposes of such discussion will be to note the distinctive resources of the method of science when used in this area, and to mark the grounds for its superiority over earlier or alternative methods such as those of myth and uncontrolled speculation generally. It is the adoption of the responsible, self-corrective method of science in cosmology which enables us to see how that discipline is able to avoid the two pitfalls that have perennially dogged its career, namely, dogmatism and scepticism. At the same time we find, in turning to the accounts of the universe that are produced in the name of scientific cosmology—particularly such as have burst upon the intellectual scene with renewed vitality in recent years—that these are not always accompanied in the minds of investigators in the field by a unanimous sense of what the method of science involves or requires. Methodological controversy, and not merely differences over questions of a purely technical character, is a notable feature of the disagreements that have come to identify various "schools." By way of illustration, there is to be found in certain quarters at the present time the revival of a philosophy of rationalism, one calling for the employment of special "deductive" methods in cosmology admittedly different from those at work in other areas of science. It will be the purpose of the following analysis not only to examine what the use of "ordinary" methods of science *does* involve in cosmology and to argue their sufficiency, but to show as well the extent to which such rationalist claims suffer from certain serious faults. (2) As a second special objective, it will be of interest to explore the way in which questions relating to the spatial and temporal extent of the universe—questions having to do with its finitude or infinity—are treated in scientific cosmologies, and what status must be assigned to such answers as are accepted for these crucial questions. That it is necessary to see such efforts as parts of the enterprise of *theory* construction, in a special sense of that term, and not in any simple sense as matters for empirical determination, will be one of the major points to be stressed.

Chapter 2

The Development of Cosmology from Myth to Science

1. Kinds and Stages of Progress in Cosmology

FROM ONE point of view, it might seem plausible to maintain, as some indeed have, that scientific cosmology is so recent a product of intellectual history that all efforts to trace it farther back than the second decade of the present century would be futile. For only since then have the resources, both observational and theoretical, become available on a scale and with a degree of technical refinement which have no equal in earlier times. The discovery of the "realm of the nebulae"—that vast collection of galaxies stretching far beyond the confines of the star-system to which our own sun and its planetary system belong—and the use of relativity theory as an important base for constructing models of the universe as a whole, together usher in, it may be said, for the first time a sufficiently powerful attack on this most ancient and formidable of all problems. They justify us in identifying these efforts as uniquely entitled to the adjective "scientific."

Now while there is no need to deny or minimize the overwhelmingly important character of these recent researches, yet a more sober regard for the historical dimension will recognize in these efforts but the latest stage of a long development. Some of those efforts that antedated the present ones were in method and spirit just as much entitled to the adjective "scientific." Indeed the very importance of these latest advances is best appreciated when seen against the background that earlier contributions made.

There are two components in all scientific cosmologies— the data provided by observational astronomy concerning what are taken to be the major constituents of the universe, and the interpretational scheme invoked to make such data intelligible. To speak of "progress" in cosmology and to look for signs of it even in the past requires, however, that we apply rather different criteria to each of these sides of the subject. On an observational level, one may speak of growth as taking the form of "receding horizons" and "increasing detail." With the aid of ever-improving instruments, greater distances can be plumbed and greater detail, both in the number and identifying traits of objects, distinguished. This in a sense is "one-line" progress. The data verified and accumulated tend to remain in the storehouse of established fact. On the level of interpretation or theory, however, the story is an altogether different one. The role of theories in science is an instrumental one; they serve to connect in an intelligible order the facts established through observation, and they help to guide inquiry in the discovery of fresh facts. Their life, however, is precarious. With the advance of knowledge on the level of insight, theories either die or become absorbed in new ones. The discovery of new conceptual patterns may resemble older ones only slightly even when they are used to provide a more adequate interpretation of facts already known. In other cases, new theories are called into existence where discovery of fresh facts are inadequately represented by older schemes of thought. On the level of theory there is no "single-line" progress. There is no sense in speaking here of a steadily increasing disclosure of the "true" structure of things, that is, by a successive refinement of the *same* pattern of ideas. While it is the task of theory to keep pace with and, if possible, anticipate the advances of observation, such growth as cosmology can make is a double-tracked one; on an observational level this consists in finding out, more and more, of what is *there* to be found, and on an interpretative level, in trying to make sense of what is thus found. There is no *a priori* reason to believe, however, that in cosmology, any more than in any other domain of science, there is any limit to be anticipated on either level—to the discovery of facts or their understanding. Thus, what is referred to as "the universe," far from having a constant and univocal sense, reflects, throughout the history of cosmology, shifts in meaning that are best understood in terms of the contributions made by the growing resources of observation and the changing perspectives of theory. A knowledge of such history supplements and confirms what logical analysis will in any case show, that what we call "the universe" is the changing complement of ever-more adequate but always incomplete knowledge.

The history of cosmology, when looked at in wide perspective, reveals four main stages in its growth. The first is marked by the gradual changeover from the use of myth, which makes the world intelligible in dramatic and anthropomorphic terms, to an approach which looks for answers in the use of physical and mathematical ideas. Here, the contributions of the Pre-Socratics, particularly the early Ionians, the Pythagoreans and the Atomists are of seminal importance. The second stage was reached with the development of the classic and well-rounded Aristotelian cosmology of a finite and geometric universe. Belonging to this conception of the world was a relatively detailed fund of astronomic observation, a comprehensive and consistent scheme of (qualitative) physical ideas, and an ingenious mathematical (geometric) interpretation of planetary motions. It remained the orthodox cosmology up to its overthrow with the beginnings of modern science. The third main stage is one which stretched from the immediate changes in outlook brought about by the Copernican revolution to the second decade of the present century. On its astronomic side this period is marked by the growing emphasis given to the "universe of stars" as distinguished from the dominating importance given in earlier schemes to the planetary system. On the physical side, the great synthesis accomplished by Newton for this era is important for cosmology not because of any satisfactory solution it was able to offer to its problems (as contrasted with the enormous success it did have in the more restricted regions of the mechanics of the solar system) but because it pointed the way for the first time to the use of a mathematically formulated dynamics in its relevance to the cosmologic problem. While no stable cosmology emerges in this third period, at least of a kind which can match the long orthodoxy of the Aristotelian scheme, its importance lies in being, as it were, a transitional phase between the breakup of that older scheme and the emergence in our own day (the fourth and current phase) of a renewed attack on the cosmologic problem that is more hopeful of positive results. This most recent period is marked, on the one hand, by the fact that it consolidates and expands earlier observational discoveries with a confirmed break-through into the "universe of nebulae." In replacing the conception of a "universe of stars" it provides a new and enlarged field for cosmologic interpretation. The introduction, too, of powerful theoretic tools, particularly in the form of relativity conceptions, is one which opens up new vistas of comprehension not dreamed of in classical physics. Again, no stable cosmology on these broadened horizons is yet in sight. But great advances have already been made and their significance from a logical point of view needs to be estimated not only in the light of earlier efforts but in terms also of the *kind* of successes which it would be reasonable as a matter of principle to expect.

2. Cosmogonical Myths: The Enuma Elish

The scientific approach to cosmology as distinguished from that of myth and uncontrolled speculation generally, has at least two marks by which it may be identified. In the first place, the subject matter which it takes for exploration, the materials upon whose foundation it would establish and test its theoretical constructions is provided by the refined products of observational astronomy. Secondly, the theoretical interpretations it comes to favor at one time or another are based on ideas of a mathematical and physical sort.

However, just as no sharp division can be made, historically, between the use of myth as the basis of cosmologic speculation and the point at which metaphysical and theologic schemes evolved out of these and became established as independent ventures, divergent from that of science, so in searching for the "first" evidences of a scientific attack of cosmologic problems, it is not possible to readily distinguish these in their earliest forms from their antecedents in myth. The fact is that primitive mythology lingers on in one form or another in the early career of science and, in the case of the efforts made in cosmologic speculation, determines the very pattern, in a broad sense, which these proto-scientific schemes exhibit. This can be seen most readily in the case of the cosmologic speculations of the sixth and fifth centuries B.C. "Pre-Socratics" commonly taken as the "first" philosophers, but who may be regarded with equal justice, at least in some cases, from the point of view of our present interest, as the first "scientific" cosmologists.

One evidence of the influence of myth upon these earliest instances of "scientific" thought is to be found in the interest in formulating a complete cosmogony which would show how from some primordial state an ordered world arose and underwent successive differentiations of an astronomic, geographic and meteorologic kind, culminating ultimately, in the emergence of living things and human society. It is this theme rather than the common claim of histories of philosophy that these earliest thinkers were primarily metaphysicians seeking some underlying real substance in terms of which the changing and manifold world of appearances could be understood, which enables us to grasp the common pattern in their thought and in many cases to reconstruct its sequence when all that have come down to us are scattered fragments of their sayings. This may be seen clearly (largely through the recent brilliant work of Cornford,) in the case, for example, of Anaximander, the thinker of the sixth century B.C., who together with Thales and Anaximenes belongs to the earliest of Ionian schools.[1]

Let us recall briefly something of the nature of those myths which provided the background against which the achievement of Anaximander and other Pre-Socratics can be judged. The form that is virtually universal in early myths such as the one to be found in Hesiod's Theogony, with which Anaximander was undoubtedly familiar, is that of a cosmogony. Man's first ventures in making the world intelligible were in the form of stories he told to himself that in effect began with "Once upon a time . . ." or "In the beginning. . . ." Stories about the cosmos were essentially "histories" describing the way in which the world originated and developed.

[1] Cf. F. M. Cornford, *Principium Sapientiae* (Cambridge University Press, 1952) Part 2.

The interpretations of the manner of origination of the world consisted in the extended use given to ideas derived from certain familiar facts of experience. One was based upon the fact of making something by hand, or *craftsmanship;* another was based on the fact of *birth and growth* or the phenomenon of organisms developing from seeds or eggs; a third was based on the *evocative power of authoritative command.* Each served as the basis for complex cosmogonies. On a quite primitive level one finds the idea of some god or hero who dismembers the body of some animal, a "dragon" for example, and out of it makes or carves a world. (Here, by the way, is to be found the original etymological meaning of the Hebrew verb "to create," *bará,* in *Genesis;* it points to the meaning "to cut" or "to carve.") The use of biological concepts is illustrated in the idea of Orphic cosmogony that the world developed out of a primordial World-Egg. Closely related, is the appeal to sexual categories as a model of cosmic genesis. From this point of view the cosmogonic process may be described in typical fashion as the outcome of the union of male and female, of Heaven and Earth. Such primeval pairs occur in the figures of *Keb* the Earth-God and *Nut* the Sky-Goddess in Egyptian mythology, or *Gaia* and *Ouranos* in the Greek version. Finally, there is the whole group of cosmogonical myths which appeal to the power of thought, speech, or will to bring forth a cosmos. To this class belongs the original conception of the "cosmos" in Greek thought which is rooted in the idea of universal Law and Destiny. Tradition generally assigns to Pythagoras the first use of the term "cosmos" as signifying the order to be found in the universe as a whole.[2] Behind this usage, however, there is already in myth the idea of a world whose order has been set up by some powerful authority. The original meaning of "cosmos" stands at first simply for the fact of a community of human beings living under law.[3] It is this idea, if not the word itself, of which myth makes use when it reads into the very constitution of the world the deliberate legal apportionment of the necessary and just bounds which no being may dare to transgress. Thus in Greek myth, as in earlier Babylonian accounts, the allotment of the provinces of heaven, sea and earth to several gods (e.g., the three

---

[2] Diogenes Laertius, *Lives of Eminent Philosophers*, viii, 48.
[3] Cf. Jaeger, W., *Paideia* (Oxford, 1943) vol. 1, 155; Cornford, F. M., *From Religion to Philosophy* (London, 1912) chs. 1, 2.

sons of Kronos) provides on a cosmic scale the parallel to the power possessed by an earthly king in apportioning major though restricted domains to his chief subordinates.

The *Enuma Elish*,[4] the Babylonian Epic of Creation composed in the second millennium B.C., is particularly interesting not only because it provides an earlier version of some of the things to be found in Hesiod as well as in the *Book of Genesis*, but also for the way it exhibits in striking fashion the combined use of a variety of images for interpreting the creation of the universe. Ideas originally suggested by the local geography of the Mesopotamian valley, the biological fact of sexual generation, the social phenomenon of authority vested in a council or a king, the making of articles of use by butchering an animal, all of these are elevated onto a cosmic scale and made to serve the purpose of explaining the genesis of the world and its structure. This combination of a multitude of images may be taken as pointing to successive layers of composition belonging to different sources and stages in cultural experience. It also explains the apparent lack of consistency in the account as a whole insofar as it provides several different versions of the same event, for example the creation of the sky and earth. This is described on one occasion as a matter of transforming the carcass of an animal, and on another, as the separating of the primeval parents from their original "unity" as locked in embrace by one of the gods—their child. (It is this latter event, the separating of the parents, sky and earth, leaving a "gap" between them, which is mirrored in Hesiod's identification of "chaos" as the beginning of all things. For "chaos" meant for him not some form of primordial disorder, but simply a "yawning gap" which might be described as empty or as filled with air.) Similarly, some of the main figures in myth represent a fusion of several different roles. Tiamat, for example, represents the salt water ocean whose waters were mingled in the primordial watery chaos with those of the "sweet waters" represented by Apsu. Together, so the myth tells us, these waters existed before land or sky or even the gods came into existence. Tiamat, however, is also the female principle, the mother who gives birth to a

---

[4] Cf. Thorkild Jacobsen "Mesopotamia," ch. 5 in H. Frankford and others, *The Intellectual Adventure of Ancient Man* (Chicago, 1946) reprinted in M. K. Munitz, *Theories of the Universe* (Free Press, Chicago, 1957), Chap. I, Sec. 1.

brood of gods. These gods, as in the story of Hesiod, revolt against the parents, one of them, Ea, killing the father Apsu. In the rest of the story Tiamat appears as the monstrous dragon against whom Marduk, the son of Ea, goes forth to do battle. Marduk is himself a complex symbol, standing in one instance for the power of the wind and storm who is able to make the waters separate thereby causing dry land to appear, or the power that is able to keep sky and earth apart "like the two sides of an inflated bag." In subduing Tiamat, Marduk as a hunter or fisherman cuts up her body into two parts to form the sky and the earth. In the sky as thus set up, Marduk (who is also vested with the supreme power of command and magic, as befits a kingly ruler) then establishes the constellations of stars whose rising and settings serve to distinguish the periods of time and help men in their calendar reckonings. The planets and the moon are similarly fixed by decree in their rounds and duties.

In all early cosmogonies, of which the foregoing is a particularly clear example, we see how myth constitutes a dramatic interpretation of the world in terms borrowed from the field of familiar human experience. Such images and their associated beliefs are not adopted as a matter of conscious allegory cloaking literal thought. It is rather the very form in which thought is expressed, since no other is available with which it can be compared or of which in turn it might serve as symbol. Such projection or extension of biological, psychological or cultural terms for purposes of an account of the cosmos are, for myth-making intelligence, not even recognized as such. Indeed the need for a distinction between living and non-living and the restriction of the terms descriptive of each to their appropriate domains, would hardly be understood on this level of thought. Man, society, and the universe form an unbroken unity such that a common storehouse of concepts may be used throughout. It is only later, more sophisticated thought that undertakes to judge by means of its own standards the relative inadequacy and naïveté of such myths. Originally, however, these were not created or adopted in the face of available alternatives. They represent rather the honest and serious effort at understanding the world in such terms and on the basis of such experiences as were to be had. The vigor and imaginativeness which such products embody, moreover, were precisely the qualities necessary for the emergence of self-critical philosophy and science.

3. Proto-Scientific Cosmologies: Anaximander

That the ordered world as we know it is not everlasting but arose in some fashion from an earlier primordial state, is for Anaximander a belief which is not questioned but rather taken over from mythology. In contrast to the manner of treatment of the latter, however, the interpretation given of what this primordial state was, as indeed the whole subsequent description of the emergence of an ordered world (insofar as his account can be pieced together at all) is one which is presented in terms which are virtually divested of all metaphoric or dramatic meaning. Instead of appealing to the images of craftsmanship, sex, or divine decree, Anaximander reinterprets, while at the same time retaining, basically the same pattern of cosmogonical development that is to be found in the Babylonian myth as this had already been partly transformed in the Greek version of Hesiod's *Theogony*. The terms in which Anaximander describes the pattern of genesis is now wholly physical in character. And this, despite the obviously primitive quality of his account, is what makes it important. With Anaximander the imagery of cutting up the body of a dragon, or that of a primeval pair locked in embrace until separated by their offspring, is dropped, but the idea of a primordial unity out of which an ordered world arose by differentiation is retained. The primordial unity is called by Anaximander "the Boundless." It is a name at once for the original state of things and for the underlying imperishable material source out of which a world arose or a succession of worlds will continue to arise. The use of the term "Boundless" need not be taken as referring to that which is quantitatively infinite or qualitatively indeterminate but simply to that which originally has no internal lines of distinction or boundaries. The process of differentiation is referred to as a "separation of opposites"; the combination and mixture of these opposites, as the analogue of sexual union, is then invoked to explain the coming into existence of the manifold objects of the world. The "separating out of opposites," paralleling the separation of Heaven and Earth in myth, becomes for Anaximander a separating out from the Boundless of the broad regions of the Hot and the Cold. The Hot is said to gradually form an incandescent mass that turns into visible fire while surrounding a core of the Cold, and where the latter undergoes further differentiation to yield another pair of opposites, the Wet and the Dry. Dry land begins to emerge from the initial watery mass by the action of the warming "heavenly" fire. The individual heavenly bodies now come into existence —the sun, moon, and stars as rings of fire enclosed in tubes of dark mist or air, and visible to us singly only at "breathing holes" like those of a flute. (Eclipses are explained as due to the blocking of these holes, the same type of occurrence that accounts for the waxing and waning of the moon.) For Anaximander the "rings" of the fixed stars and planets are closest to the earth while beyond these in order are those of the moon and finally the sun. At the center of the entire system is the Earth which "is freely suspended and not held in its place by anything, but staying there because of its equal distance from everything." Its shape is said to be cylindrical like a drum on one of whose surfaces we are situated.

The subsequent history of Pre-Socratic cosmology is one in which the advance marked by Anaximander is carried forward in many different directions. Of these efforts, the most significant were those made by the Pythagoreans and the Atomists, the one giving emphasis primarily to mathematical ideas, the other primarily to physical ones. The cosmology of Plato as expressed in the *Timaeus,* in reverting to the technique of myth, represents, on the whole, a fateful step backward in the history of the subject. The re-introduction of teleologic ideas in conjunction with a degenerate form of Pythagorean number-mysticism was associated with a deliberate rejection of the promising and prophetic ideas of the atomistic materialists. Aristotle's achievement, such as it was in cosmology, consisted, not in breaking away wholly from the Platonic framework—which he surely did not—but in giving greater prominence to purely physical ideas (even though these were of a type heavily influenced by biologic models) in place of the myth-sanctioned teleologic ones of the *Timaeus.* Aristotle thus accomplished on a fairly elaborate scale with respect to Plato the same sort of thing that Anaximander had achieved with respect to Hesiod.

4. The Classical World-View: Aristotle

For Aristotle, to whom we may next turn, therefore, as marking another major stage in cosmological speculation, the universe is a particular, material thing, a determinate whole composed of all natural perceptible body included within the extreme circumference of the heaven.[5] The distribution of matter and motion in this universe—the types of bodies that compose it and their corresponding motions—constitutes the basic problem, the answers to which specify the properties which the universe possesses as a whole. The qualitative theory of matter and motion invoked, rests on the assumed fundamental distinction between circular and rectilinear motion. Since the four basic elements, earth, water, air, and fire, of which all compound bodies are made up have natural rectilinear motions, upward or downward, it follows for Aristotle that there must be some body which has a simple natural motion that is circular. Since, moreover, the circular is considered perfect, whereas the straight line is not, it will be naturally prior to the others. A fifth element, the ether, is that of which, therefore, the heavenly bodies are composed. Circular motion, unlike the rectilinear, admits of no contraries such as 'up' or 'down' hence the heavenly bodies which possess such motion will be neither heavy nor light. For heaviness and lightness are respectively defined in terms of that which naturally has a downward motion toward the center and that which naturally has an upward motion away from the center.[6] Again, since contrariety does not obtain for ethereal bodies, they will be eternal, and this for Aristotle means that they will be ungenerated and indestructible, free from increase and alteration. Indeed the unchangeability of the heavenly bodies is confirmed for us according to Aristotle by traditional experience which reveals no change "either in the whole scheme of the outermost heaven or in any of its proper parts." [7] The sublunar or terrestrial region is the abode of the four elements where, by contrast, change and transformation are prevalent. From such a theory of matter and motion, the properties which the universe as a whole possesses logically follow. The universe must be a finite one spatially, since no infinite body can traverse in a finite time a complete circle. The diurnal motion of the sun and the stars provides a stubborn and obvious fact with which the supposition of an infinite body for the universe would not be consistent. "Our eyes tell us

[5] De Caelo, 278b 3, 9, 23.
[6] ibid. 269b 30.
[7] ibid. 270b 15.

that the heavens revolve in a circle." [8] Again, in opposition to the doctrine of a plurality of worlds (such as the Atomists maintained) Aristotle insists there is but one universe, i.e., one planetary system and outermost shell of fixed stars, and that this entire system contains all existent material bodies. To suppose otherwise would be to contradict the theory of matter and motion assumed, since bodies, for example, that naturally move downward to a given center could not consistently and simultaneously move there if there were more than one center, as would be the case if there were more than one world. A corollary of the fact that there is but one world and that there is no body beyond its confines is that "there is also no place or void or time outside the heaven" since all of these are connected with the existence of body. The spherical shape of the universe is established on several grounds, among them the fact that only this shape permits rotation in one place since there is no space or void outside the heavens.

In order to account for the motions of the sun, moon, planets and fixed stars, Aristotle adopts the theory originally worked out by Eudoxus of Cnidus (409–356 B.C.) and modified by his pupil Callipus of Cyzicus (370–300 B.C.). This theory in its original Eudoxian form was directed to an explanation primarily of the motions of the planets but was extended to include other heavenly bodies as well. [9] It postulated, from a purely geometric point of view, a series of concentric spheres with the Earth as their common center. Assume a given planet fixed to the equator of a sphere revolving uniformly about its poles. Then in order to account for the actually observed apparent irregularities in the motions of the planets (retrogressions, variable velocities, stationary points) it is necessary to imagine the poles of the sphere carrying the planet themselves fixed to another sphere with its own poles and its own period of uniform motion and this in turn fixed to some other, and so on, until a specific combination of spheres, the resultant of different proper inclinations and proper periods of rotation, yields the required observed motions. For Eudoxus the combination of three spheres each for the sun and the moon, four for each of the

[8] ibid. 272a 5.
[9] Cf. Drever, J. L. E., History of Planetary Systems from Thales to Kepler (Cambridge, 1906) ch. iv; Heath, T. L. Aristarchus of Samos (Oxford, 1931) ch. XVI.

five planets, and one for the daily rotation of the fixed stars, twenty-seven spheres in all, was taken as sufficient. Callipus, in order to improve the accuracy with which the phenomena might be described, found it necessary to postulate two additional spheres for the sun and moon and one additional for three of the planets. Aristotle takes this basic scheme, certainly by far the best available in his day, and transforms it by making the spheres, which in the Eudoxian scheme were purely geometric, into a physical system, a series of spherical shells in contact with one another and transmitting thereby their several motions.[10] This continuous nest of shells required, however, in addition to those which Eudoxus and Callipus proposed, an additional number of "unrolling" spheres, and this gives in Aristotle's scheme a total of fifty-five.[11] The outermost shell of the universe, containing the fixed stars, performs its unceasing diurnal motion in a perfectly regular manner and is kept in motion by the single eternal Prime Mover or God operating as an object of love and desire. From the extreme outer sphere of the fixed stars with its simple motion, we pass successively to the other heavenly bodies with their more complicated motions until we reach the Earth, which, Aristotle argues, must be conceived of as relatively small, spherical in shape, motionless and occupying the absolute center of the entire universe.

The just summarized features of the Aristotelian cosmology provided the basic outlook of civilized thought until the revolutionary changes wrought by modern science caused virtually its complete abandonment. The details of Aristotle's cosmology did not, of course, remain entirely unchallenged or unmodified until the days of Copernicus. Such changes as were made, however, did not serve to completely shatter the picture it offered. The two major, significant changes from Aristotle's day to the sixteenth century affected rather the mathematical and astronomic detail, on the one side, and the theological framework for the entire scheme on the other. The first major modification consisted in the substitution of the Ptolemaic system of eccentrics and epicycles for the system of concentric spheres. The chief reason for the abandonment of the latter theory consisted in its failure to account for the great variation in the brightness of the various planets at different positions in their orbits, a variation un-

[10] *loc. cit.,* 289b-290a.
[11] *Metaphysics* 1073b-1074a, cf. Heath, *loc. cit.* ch. XVI.

accounted for by the assumption of their equal distances from the Earth. Like Eudoxus centuries before, Ptolemy looked upon his own account of the planetary motions as a purely mathematical device without pretending to invest it with mechanical reality. Just as Aristotle had materialized the scheme of Eudoxus, however, so, many of the successors of Ptolemy among the Arabs and Schoolmen, gave body to the Ptolemaic scheme by investing the eccentrics and epicycles with physical existence by taking them as belonging to a number of crystalline spheres. While the machinery was thus tremendously increased in complexity, the basic geocentric and finitist view was left untouched. The second major change in the Aristotelian scheme concerned the underlying metaphysics to which it was subordinated. The Aristotelian doctrine concerning the ungenerated and indestructible character of the universe was universally rejected during the Middle Ages as contrary to Judaeo-Christian theology, and the notion of a creation *ex nihilo* was put in its place. Thus the resultant cosmology that prevailed toward the end of the Middle Ages was a fusion of three elements: Aristotelian physics, Ptolemaic astronomy and Mosaic cosmogony.

It was this very dominance of the Aristotelian philosophy and the geocentric point of view which it incorporated that led to bypassing the significant astronomic clues in ancient thought that might have helped to overthrow this whole scheme. The germinal ideas of Heraclides of Pontus (a contemporary of Aristotle) who developed the hypothesis of the earth's daily rotation on its axis and of the revolution of Venus and Mercury around the sun as satellites; and the even more prophetic and thoroughgoing ideas of Aristarchus of Samos that contained the equivalent of a Copernican, heliocentric scheme of planetary motions, failed to gain the recognition and elaboration they deserved. Hipparchus, the greatest astronomer of antiquity, and following him, Ptolemy, adhered to the geocentric theory and thus lent their overwhelming prestige to the continuance of the entire fabric of Aristotelian cosmology.

5. The Universe of Stars in Modern Astronomy

The immediate result of the Copernican revolution was to bring about a more effective theory of planetary motion. Its wider cosmological bearings were not of special concern to

Copernicus himself who adopted on this score a conservative attitude. He simply looked now to the sun rather than to the Earth as the center of the universe. He retained the geometric scheme of epicycles and eccentrics although the complexity required by the Ptolemaic theory could now be enormously reduced. No longer was it necessary, however, to postulate the daily revolution of the sphere of the fixed stars. That there was such a sphere, though, Copernicus does not doubt, even if its distance from the Earth is increased by a tremendous amount in order to account for the absence of a stellar parallax in the Earth's traversal around the sun. With respect to the extent of this sphere of fixed stars, whether it is finite or infinite, however, Copernicus refused to commit himself. That is something, he remarks, which may be left to the philosophers to speculate about.[12] The job of the astronomer, more restricted but at the same time more hopeful of solution, is the description of what comes within the limits of observation and its mathematical articulation. In this respect Copernicus was followed by Galileo who writes that no one "has so far proved whether the universe is finite and has a shape or whether it is infinite and unbounded." [13]

But the wider, more speculative effects of the Copernican theory upon the outlook of cosmology were not slow in being felt and elaborated by others. Thomas Digges, for example, one of the early English translators and supporters of Copernicus, in 1576 undertook to make a deliberate innovation in the picture of the universe which the new astronomy seemed to make possible. Since technically it was no longer necessary to picture the universe as limited by a finite outermost sphere of fixed stars, as the older astronomy with its assumption of a fixed central Earth required, it could now be pictured as extending infinitely in all directions. Digges becomes, therefore, the "first modern astronomer of note to portray an infinite, heliocentric universe with the stars scattered at varying distances throughout infinite space." [14] What is important in this and other early speculative ventures, as well as in later more refined astronomical investigations, is the fact that they

---

[12] Cf. Johnson, F. R., *Astronomical Thought in Renaissance England* (Baltimore, 1937) 106; E. Rosen, *Three Copernican Treatises* (New York, 1939) 39.

[13] *Dialogue Concerning the Two Chief World Systems*, transl. by Stillman Drake (Univ. of Calif. Press, 1953) 319.

[14] Johnson, F. R., *op. cit.*, 164.

now exhibit a new frame of reference for thinking in cosmology. For in taking the collection of *stars* as the domain of greatest inclusiveness, as "the universe," the variety of efforts made in cosmological speculation and astronomical inquiry was now concentrated upon determining *its* structure. Indeed, once having made the sun a member of the family of stars, there was no longer strictly any need as either Copernicus or Digges felt it, to assign a unique, central position to our own sun. Does then the stellar system as a whole have a particular pattern of spatial distribution? If finite is it isolated in space or are there other comparable systems? These questions and the efforts at answering them begin to come into prominence beginning in the eighteenth century. Effectively reliable answers however, were not forthcoming until our own generation.

One early significant step in the search for a possible structure in the universe of stars was taken by Thomas Wright of Durham in 1750.[15] Instead of taking the stars as scattered at random out to infinitely large distances, Wright proposed a consideration of the familiar phenomenon of the Milky Way as providing a major clue. Already with the first use of the telescope by Galileo it had been confirmed that what otherwise might be taken as a streaky nebulosity, was in fact, in what is called the Milky Way, a vast congeries of individual stars. Wright's strikingly novel suggestion was that the whole system of stars be viewed as one enormous, finite, disc-like structure. Our own sun and its attendant planets may be considered as lying in the plane of this structure somewhere near its center. In looking out towards the top or "poles" of the arrangement, relatively few stars will be met. On the other hand, as we go towards the rim of the system, more and more stars are found, until at the rim itself what we get, due to the perspective in which this vast thickness of stars will appear to a member of the system such as ourselves, will be precisely the effect produced by the Milky Way. Along with this startling and prophetic idea Wright likewise conceives of the existence of other "universes" beyond our own, that is, stellar systems, or, as he phrases it, "creations." That this "a Plenum of Creations not unlike the known universe may be the real case" he writes "is in some degree made evident by the many cloudy spots,

---

[15] *An Original Theory or New Hypothesis of the Universe* (1750); an American edition of this work was published in Philadelphia in 1837.

just perceivable by us, as far without our starry regions in which through visibly luminous spaces, no one star or particular constituent can possibly be distinguished; those in all likelihood may be external creations, bordering upon the known one, too remote for even our telescopes to reach." [16]

To Kant, in his *Natural History and Theory of the Heavens* (1755) must go the credit for having worked out in somewhat elaborate detail (on the basis of an account of Wright's theory which he read) a conception of the Milky Way and of comparable external systems that is remarkable for its imaginative insight and reasoning power. "If a system of fixed stars," he writes, "which are related in their positions to a common plane, as we have delineated the Milky Way to be, be so far removed from us that the individual stars of which it consists are no longer sensibly distinguishable even by the telescope; if its distance has the same ratio to the distance of the stars of the Milky Way as that of the latter has to the distance of the sun; in short, if such a world of fixed stars is beheld at such an immense distance from the eye of the spectator situated outside of it, then this world will appear under a small angle as a patch of space whose figure will be circular if its plane is presented directly to the eye, and elliptical if it is seen from the side or obliquely. The feebleness of its light, its figure, and the apparent size of its diameter will clearly distinguish such a phenomenon when it is presented from all the stars that are seen single. We do not need to look long for this phenomenon among the observations of the astronomers. It has been distinctly perceived by different observers. They have been astonished at its strangeness; and it has given occasion for conjectures, sometimes to strange hypotheses, and at other times to probable conceptions which, however, were just as groundless as the former. It is the 'nebulous' stars which we refer to, or rather a species of them, which M. de Maupertuis thus describes: 'They are,' he says, 'small luminous patches, only a little more brilliant than the dark background of the heavens; they are presented in all quarters; they present the figure of ellipses more or less open; and their light is much feebler than that of any other object we can perceive in the heavens.' " Kant then rejects Maupertuis' own suggestion that these are single bodies whose shape is produced by their

[16] *op. cit.*, 143.

rotational motion; instead he argues "it is far more natural and conceivable to regard them as being not such enormous single stars but systems of many stars, whose distance presents them in such a narrow space that the light which is individually imperceptible from each of them, reaches us, on account of their immense multitude, in a pale uniform glimmer." [17] Making, moreover, a characteristic appeal to a combined use of Democritean atomism, Newtonian principles, and what is taken as a mark of God's infinite power, Kant goes still further and envisages an infinite supersystem as the outcome of the systematic inter-relation of these innumerable sub-systems. He sets up a conception of what he calls a "systematic constitution" for the universe as a whole on the analogy of what had been discovered to exist in the solar system (a single massive center surrounded by bodies moving in a common plane and held together by gravitational forces). Such a "constitution," he proposes, holds in a general way not only for our own galactic system and for other stellar systems outside our own, but for the total system that binds these together in the most inclusive way. "But what is at last the end of these systematic arrangements? Where shall creation itself cease? It is evident that in order to think of it as in proportion to the power of the Infinite Being, it must have no limits at all . . . the field of the revelation of the Divine attributes is as infinite as these attributes themselves. Eternity is not sufficient to embrace the manifestation of the Supreme Being, if it is not combined with the infinitude of space." [18]

The man who first established the study of the "universe of stars" upon a sound observational basis was William Herschel (1738–1822). Interest in cosmology, or as he calls it, "the construction of the Heavens" was from the very beginning the ruling concern to which all his tremendously detailed and numerous observational studies were directed. In this domain, he remarks in one of his earliest papers on the subject, it is of the utmost importance to keep a sane balance between observation and theory. "If we would hope to make any progress in an investigation of this delicate nature we ought to avoid two opposite extremes, of which I

---

[17] W. Hastie, *Kant's Cosmogony* (Glasgow, 1900) a translation of the above mentioned work of Kant, 61-63; reprinted in M. K. Munitz, *Theories of the Universe* (Free Press, Chicago, 1957) Chap. III, Sec. 18.
[18] *loc cit.*, 138; cf. *ibid.*, 65, 154.

can hardly say which is the most dangerous. If we indulge a fanciful imagination and build worlds of our own, we must not wonder at our going wide from the path of truth and nature; but these will vanish like the Cartesian vortices, that soon gave way when better theories were offered. On the other hand, if we add observation to observation, without attempting to draw not only certain conclusions, but also conjectural views from them, we offend against the very end of which observations ought to be made." [19] His own painstaking researches and the speculations which were guided by exact observations, underwent significant changes from the earliest efforts presented in his papers of 1784–85 to the latest in 1811.[20] Herschel started out by not only adopting the view which assigned a definite, finite and generally disclike structure to the Milky Way system, he also regarded it as a "detached nebula" whose stellar structure is comparable to that of the various (in-principle) resolvable nebulae which he believed are scattered beyond the confines of our own system. His later observations and interpretations gradually modified this view to the extent that he discovered, increasingly, the complex structure of the Milky Way itself—as not composed of "insulated" stars throughout—and the fact that some nebulae turned up which were not in all cases resolvable into stars. These latter Herschel took as made up of a "shining fluid" and "of a nature wholly unknown to us." He tried to account for the variety of telescopic objects already discovered by an ingenious hypothesis of evolutionary development, of change from diffuse nebulosities into stellar systems. At no time, however, did he venture to make any claims about the nature and extent of the total system comprising all these sub-systems.

The subsequent hisory of nebular research in the nineteenth and early twentieth centuries exhibits a shifting and complicated pattern of observation and interpretation in which the hypothesis of the existence of "island-universes" outside our own galaxy is now brought into prominence and now discarded to be replaced by the view which would absorb all

[19] "On the Construction of the Heavens" (1785) in *Collected Papers*, ed. J. L. E. Dreyer (1912) I, 223; reprinted in M. K. Munitz, *Theories of the Universe*, Chap. III, Sec. 20.
[20] Cf. F. G. W. Struve, *Etudes D'Astronomie Stellaire*, 21-50; H. Macpherson, *Modern Astronomy* (Oxford, 1926) 144ff.; C. Lubbock, *Herschel-Chronicle* (New York, 1933).

types of heavenly bodies within a single stellar system. Thus with the use of the 6-foot reflector of Lord Rosse the "island-universe" hypothesis gained favor beginning in 1849 as a result of its ability to resolve some of the so-called spiral nebulae into stars. Writing in 1852 Grant concluded: "The phenomena denominated Nebulous Stars—which seemed to Herschel to be incapable of any satisfactory explanation, except by adopting the hypothesis of a self-luminous fluid—when examined with the powerful telescopes of Lord Rosse, have been found to exhibit an aspect totally different from that which appeared to Herschel so enigmatical. In fact, the greater the optical power of the telescope with which the heavens are surveyed, the more strongly do the results tend to produce the impression that all nebulae are in reality vast aggregations of stars, which assume a nebulous aspect only because the telescope with which they are observed in each instance is not sufficiently powerful to resolve them into their constituent parts and thereby disclose their real nature." [21] Such a judgment was soon to be called into question, however, when in 1864 Huggins, with the aid of the newly invented spectroscope, discovered the bright-line or "gaseous" type spectra of several irregular and planetary-type nebulae. Opinions veered from one extreme to the other and opposition to the "island-universe" hypothesis continued down to the early part of the present century. Many tacitly assumed that even the so-called "white-nebulae" or spirals, though giving a continuous spectrum, would also be found to be inherently gaseous.[22] Moreover, apparently anomalous facts about the distribution of nebulae, for example that they appear predominantly in the regions about the galactic poles, avoiding the region of the galactic plane, were taken by many as indicative of their association with the galaxy and so part of the stellar system. The historian of astronomy Agnes Clerke, writing in 1905 summed up the opinion current at the time when she wrote: "The question whether nebulae are external galaxies hardly any longer needs discussion. It has been answered by the progress of research. No competent thinker, with the whole of the available evidence before him, can now, it is safe to say, maintain any single nebula

[21] Grant, Robert, *History of Physical Astronomy* (1852) 568.
[22] Cf. Curtis, H. B., "The Nebulae" *Handbuch der Astrophys.* (Berlin, 1933) vol. 5, 833-34.

to be a star system of co-ordinate rank with the Milky Way." [23]

Paralleling these many important yet inconclusive efforts of observational astronomy to reach some decision concerning the structure and distribution of the stars or systems of stars of which the universe was taken to be composed, the contribution of physical theory in the form of the universally accepted Newtonian mechanics towards an understanding of the cosmologic problem presents an equally chequered career. While it does not at any point provide an adequate or lasting solution, yet in grappling with the problem, many suggestive proposals are elicited, ones that in some cases pave the way for the better equipped attack which relativity theory was later able to make.

The assumption that the total amount of matter in the infinitely extended Euclidean space of the universe is concentrated in a finite region, was rejected by Newton on the ground that "the matter on the outside of this space would by its gravity tend toward all the matter on the inside and, by consequence, fall down into the middle of the whole space and there compose one great spherical mass." [24] (Einstein when regarding the cosmologic problem from a broad Newtonian point of view, rejects the same conception of a "finite (stellar) island in an infinite ocean of space" because "it leads to the result that the light emitted by the stars and also individual stars in the stellar system are perpetually passing out into infinite space, never to return, and without ever again coming into interaction with other objects of nature. Such a finite material universe would be destined to become gradually but systematically impoverished." [25]) However, the contrary assumption, one that might seem at first glance more attractive on the classical basis, that the universe as a whole be conceived as consisting of an infinite amount of (stellar) matter spread out uniformly in infinite Euclidean space, with a finite density throughout, presents equally serious difficulties. One of the most serious of these was pointed out by Seeliger in 1895. If we assume the Newtonian inverse-square law of gravitation, calculation shows that forces acting at any point upon some body as the effect of an infinite num-

[23] *The System of the Stars* (1905) 349; cf. same author's *History of Astronomy During the 19th Century* (London, 1902) 422.
[24] *First Letter to Bentley*, cf. M. K. Munitz, *Theories of the Universe* (Free Press, Chicago, Ill. 1957) Chap. III Sec. 15.
[25] *Relativity* (London, 15th ed. 1954) ch. 30, 106.

ber of masses yields no finite value but only infinite and indeterminate ones.[26] Seeliger's own solution, that Newton's law be modified in such a way that the force of gravitational attraction be conceived as falling off for bodies at very great distances from one another, was simply an *ad hoc* explanation and did not find favor. Earlier in that century (1826) another objection, one made by Olbers, against the conception of a uniformly distributed infinite amount of stellar matter (all with approximately the same degree of brightness) scattered throughout infinite Euclidean space, was that under such conditions the sky should be infinitely bright, which, of course, contradicts the relatively slight amount of radiation actually observed. Once more various *ad hoc* ways of meeting this objection were proposed, among them one by Olbers that there is an absorbing, tenuous gas in interstellar space which cuts down the radiation received.

One important theoretical line of attack which showed how the foregoing objections might be overcome while still retaining the Newtonian system of mechanics and the use of Euclidean geometry, was made by the Swedish astronomer Charlier who, on the basis of suggestions of the eighteenth century thinker, Lambert, proposed the idea of a hierarchic structure of the universe as a whole. According to this model, the universe is built up of an infinite sequence of systems of increasing inclusiveness. Thus, let a single star such as our sun be one instance of a system of lowest rank; then let all such systems (stars) form an order of next higher rank, that is, a galaxy or star-system, and let all such systems of stars in turn (galaxies) form a system of still higher rank, which again will be but a member along with all comparable systems of equal rank in a system of still higher rank, and so on. Charlier showed how under these conditions the objections of

[26] Einstein's proof of this is as follows: "According to the theory of Newton, the number of "lines of force" which come from infinity and terminate in a mass m is proportional to the mass m. If, on the average, the mass-density p_0 is constant throughout the universe, then a sphere of volume V will enclose the average mass p_0V. Thus the number of lines of force passing through the surface F of the sphere into its interior is proportional to p_0V. For unit area of the surface of the sphere the number of lines of force which enters the sphere is thus proportional to $p_0 \dfrac{V}{F}$ or to p_0 R. Hence the intensity of the field at the surface would ultimately become infinite with increasing radius R of the sphere, which is impossible." *Relativity*, 106.

both Seeliger and Olbers may be removed.[27] The need for such a "hierarchic" solution, however, became less keenly felt with the continuing advance of observation and theory. In the first place there is no present observational warrant for recognizing any order of astronomical bodies higher than the system of galaxies themselves, and for these a *uniform* distribution is generally accepted. Also the use of relativity theory, including the conception of an expanding universe, has lessened the appeal which Charlier's hierarchic approach had, founded as it is on a Newtonian basis, Euclidean geometry, and the absence of an overall large-scale systematic motions for the various populations of astronomic bodies.

More recent investigations initiated by Milne and McCrea (in 1934) have shown, however, that even apart from the appeal to relativity conceptions, various models of the universe founded on classical Newtonian ideas may be consistently constructed, provided that the assumption is surrendered which denies any large-scale motions to the matter of the universe as a whole. Such "Newtonian" models, while admittedly not as adequate as those developed in terms of the more refined theoretical concepts now available, are nevertheless instructive and useful in suggesting analogues for the latter.[28]

6. Nebulae and Relativity

Improved observational resources in our own day finally served to establish a satisfactory solution to the earlier hesitancies about the existence of external galaxies. In 1924 Hubble showed through the use of improved criteria of distance, particularly the identification of Cepheid variables as a basis for such measurements, that some nebulae are systems of stars and are indeed extra-galactic. A fundamental scheme of classification came to be accepted which recognized that a distinction must be made between the planetary or gaseous "nebulae" which are parts of our own galactic system and the truly extra-galactic nebulae. The latter consist of a variety of types, including the so-called "regular" group, those with a symmetrical structure, the compact globular systems and the

[27] Charlier, C. V. I. "How an Infinite World May Be Built Up" *Arkiv fur Matematik Astronomi och Fysik*, XVI, no, 22 (1922).
[28] On the subject of "Newtonian" cosmology see Bondi, H., *Cosmology*, Chap. IX, McVittie, G. C., "Relativistic and Newtonian Cosmology" *Astron. Journ.* 59 (1954) 173ff.

open-armed spirals.[29] The solution to the earlier-stressed anomalies in the apparent distribution of nebulae, the concentration of nebulae in the region of the galactic poles and their absence in that of the galactic plane is explained as due to the heavy obscuration by interstellar matter *within* our galaxy, most prevalent in the galactic plane, and the relative thinning out or absence of such obscuring matter as we mount towards the poles of our own system. Similarly with a proper distinction between those nebulae that are intra-galactic and are composed of gas and dust and those which being extra-galactic are, in many instances, resolvable into stars, the differences in spectral data become readily intelligible.

The present observable region of the universe, then, is constituted of vast swarms of nebulae or galaxies. Two important observational results have thus far emerged from the general reconnaissance of this region accomplished by our most powerful telescopes. The first is the fact that, generally speaking, and on the basis of the sample so far observed, the distribution of the nebulae in space is uniform. If the volume of space considered is sufficiently large, there does not appear to be any thinning out, any clustering or any special grouping of the nebulae into some super-system. A second, more startling and at the same time more perplexing phenomenon is that connected with the red-shift in the spectra of the nebulae. This has been formulated by Hubble in a way which established a generally linear relation between the red-shift and the apparent magnitude (or distance) of the nebulae. When the red-shift is interpreted as a Doppler effect, it is taken as a mark of recessional motion and the interpretation of Hubble's formula leads to the conception of an "expanding universe." The more distant the nebulae, the faster they are receding from the point of observation.

Theoretical cosmology in the hands of mathematical physicists has kept pace with observational advances by offering a variety of "models" for the systematic ordering and incorporation of the fruits of observation. Beginning with the groundbreaking paper by Einstein in 1917, where an extension was made of relativistic principles to the cosmologic domain, a variety of further attempts were made along "classical" relativistic lines by Einstein himself, de Sitter, Robertson, McVittie, Tolman, Lemaître and others, while somewhat dif-

[29] Cf. Hubble, E., *The Realm of the Nebulae* (Yale, 1936).

ferent modes of approach were subsequently opened in several directions by Milne, Eddington, Bondi and Gold, and Hoyle. While nothing in the way of a stable or wholly acceptable theory has yet made its appearance, nevertheless fruitful and promising lines of research have been inaugurated.[30] We shall return in what follows to a more detailed account of certain aspects of these more recent investigations.

Chapter 3

Some Questions of Logic in Cosmology

PROGRESS IN the field of cosmology can take place not only where a deliberate preference for the method of science is operative but where at the same time some explicit and critical comprhension exists of what such a method involves with respect to its capacities and limits. In cosmology no less than in other areas, however, such critical comprehension is not always evident in the same degree, nor is complete agreement even about fundamentals present among the very individuals engaged in this pursuit. Despite, therefore, the overall advances in scientific insight already had or likely to come, it need occasion no surprise if we find in a discipline at once so ambitious in scope, yet actually in a comparative sense only in its scientific infancy, a certain amount of hesitancy, even obscurity and underlying conflict in the conceptions adopted by some as to its proper methods. One reason for this is the fact that scientists themselves are products of an intellectual and cultural heritage with diverse components. Another is the fact that in a subject as difficult as the present one, the rules that govern or should govern its procedures call for constant re-examination. Such rules are stipulations or conventions as

[30] Cf. H. P. Robertson, "Relativistic Cosmology" *Rev. Mod. Phys.*, 5, 1933, 62-90; R. C. Tolman, *Relativity, Thermodynamics and Cosmology* (Oxford, 1934); O. Heckmann, *Theorien der Kosmolgie* (Berlin, 1942); G. C. McVittie, *Cosomological Theory* (London, 1949); G. J. Whitrow, *The Structure of the Universe* (London, Hutchinson's Univ. Library); H. Bondi, *Cosmology* (Cambridge, 1952); E. A. Milne, *Relativity, Gravitation and World Structure* (Oxford, 1935) and the same author's *Kinematic Relativity* (Oxford, 1948); W. H. McCrea "Cosmology" in *Reports on Progress in Physics* (Physical Society, London), XVI (1953), 321-363.

to how such inquiries are to be conducted in order to be successful. The determination of what constitutes success, however, cannot be established *a priori*, once for all. It must be determined in the light of the actual outcomes reached through the use of a variety of techniques and procedures. And this is something about which differences of opinion are likely to arise in a perfectly natural way. It is also a situation which invites logical analysis and provides indeed a fertile field for its exercise.

1. Deduction or Extrapolation? Space and Time

Let us consider briefly some examples of questions primarily of a logical kind, provocative of controversy, which have arisen in the carrying out of a program of cosmological inquiry.

Traditionally, as well as at the present time, questions that have occupied the forefront of interest in characterizing the universe as a whole are those which relate to its spatial and temporal dimensions. With respect to its spatial extent, is the universe finite or infinite? When considered from the point of view of its temporal career, did the universe have a beginning or was it always in existence? Will it ever come to an end or is it in some sense indestructible? It would, of course, be both naive and erroneous to assume that in asking these questions one finds in every case either clear or univocal meanings attached to the terms employed. The fact is that both in the asking of the questions as well as in the efforts at answering them, or for that matter even in refusing, as in Kant's critical philosophy or other forms of scepticism, to undertake or expect answers, one finds certain assumptions made and concepts employed which together constitute some particular theory, whether crude or sophisticated, as to the nature of space and time, as well as some particular conception of the physical universe to which these are applied.

A theory of space and time is thus an integral aspect of cosmology. In the establishment of such a theory there are two broad and interdependent tasks that need to be performed. One is technical or scientific, the other is philosophic. Looked at in wide perspective, the terms "space" and "time" may serve as convenient designations for what is encompassed in certain types of calculation and measurement. One set of problems connected with the making of such calculations and measurements concerns the definition and exact formulation of various species of quantities and ordering devices that are to be used for the symbolic representation of these operations, their expression in certain abstract mathematical laws, the manner in which these may be systematically arranged, and the conditions specified for the symbolic transformations of the concepts involved. This is the task of pure mathematics. The application of these calculi embodying the results of such mathematical investigations, their incorporation within the framework of a total physical theory, is the task of empirical science in its use of these abstract languages. In carrying through this enterprise, however, there is inevitably present a controlling set of assumptions as to the manner and significance of what is accomplished. To make explicit, critically examine, and justify the use of these assumptions is the task of philosophy. It is concerned with the understanding of the relation of mathematics to empirical science and of scientific knowledge to the subject matter which it seeks to express. To have some reasoned account of these matters is to have a philosophy of science, a theory of knowledge. In the pursuit of cosmological inquiries, the importance of philosophical considerations as they bear on the adoption and use of theories of space and time is to be noted for contemporary discussions no less than for those in the past.

Some illustrations of specific questions, at bottom of a logical character, that have been actively discussed in recent literature are the following: Is the determination of the curvature of space something that hinges upon empirical data, or is every application of a geometry arbitrary and at the disposal of the investigator, a matter of convention? How, moreover, are space-measures connected with time-measures? Does the use of a four-dimensional geometric language as in Einsteinian relativity adequately express their relationship or is it the case that in some clearly demonstrable manner time-measures are the more fundamental? If the latter, for example, what is involved in such priority? When reference is made to a "beginning" of the time sequence or to "creation," what precisely do these terms signify? In the construction of cosmologic theories is it necessary to proceed through the application and extrapolation of such physical laws as have been verified in some limited domain, or can independent, original principles be formulated for the universe as a whole which have these other laws as special consequences? Is there some limit to the rational deducibility of empirical laws such that at some point reference must be made to the given for the determination of particular quantitative functions, or is it possible for all laws of physics to be logically derived from an intuitively established axiomatic base? Is it possible in the latter case to find this base in the very process and possibility of acquiring knowledge for some conscious organism? Is it the case that there are some meaningful questions about the universe as a whole which refer to matters that indeed transcend the bounds of observational verification or the applicability of physical laws? How indeed is the range of that which is to be included within the "universe" to be determined?

The foregoing questions may be conveniently grouped and reduced to the following two sets, and it is to them that the following pages will be devoted.

(1) Is it the case that in studying the universe, one is compelled by virtue of the distinctive nature of the subject matter to adopt a method or an attitude about the very possibility of attaining knowledge here which are in important respects different from those adopted in the study of other subject matters? What, in any case, is involved in "ordinary" scientific method and what type of results are to be found or expected when it is employed in the study of the universe?

(2) When we seek to determine how the universe is to be described spatially or temporally, particularly with respect to the questions of its finitude or infinity, what are we to understand by what is accomplished in such accounts? Specifically, the question of the "objectivity" or "reality" of such accounts has to be examined. Does it make sense, and if so, what is it to say that the universe has a finite "age" or had a "beginning" or is "finite" or "infinite" in its spatial extent?

2. Importance of Philosophical Analysis

The issues summarized above have at once a classic ring, recalling numerous discussions in the past; they also have a distinctively contemporaneous importance, born as they are in the friction of controversy over rival theories struggling to establish themselves. It should be clear, however, that what is involved concerns neither the acceptability of particular observational data nor the mathematical adequacy of certain deductive chains of reasoning. The issues rather concern the views held with respect to the relations between the data and the mathematics, in short, they involve ultimately fundamental principles in the theory of knowledge. The primary and constructive task of philosophy (whether performed by "scientists" or "philosophers" is simply a matter of names) is that of clarifying the points at issue and making critical evaluations of rival positions. This task, however, is one which it would be fatal to regard as either "optional," "a matter of taste," or of "private opinion." Rather it is crucially essential to the pursuit and advancement of the discipline of cosmology. Without it, the accumulation of observational materials and the refinement of mathematical techniques lose their relevance. Philosophical questions are not outside of or superimposed on scientific ones; one cannot decide not to include them within the scientific framework, for they are already there.

Chapter 4

The Nature of Scientific Theory

THE GOAL of inquiry in empirical science, generally, is the establishment of a theory which enables us to understand some domain of observationally ascertained fact. While it is frequently said that the main object of science is the discovery of *laws* in the sense of "inductively warranted descriptive generalizations of observed fact," this is, at least for advanced sciences, best seen as subsidiary to the interest of formulating *theories*. For while it is true that "laws" in the sense just indicated may be said on one level to "explain" individual occurrences or instances that can be subsumed under them, it is theory and not still wider or more inclusive generalizations of fact which explains such "laws" themselves. In mathematical physics, theories are contained in a fundamental set of equations together with such accompanying commentary as is necessary to explain their use. Classic and familiar examples of such theories are Newton's Theory of Mechanics as contained in his so-called Laws of Motion, Maxwell's Theory of the Electromagnetic Field, and Einstein's Theory of Relativity in its special and general forms, the last expressed in the form of a set of general field equations.

1. Functions of Theory

Theories serve a variety of functions of which three may be singled out for discussion because of their importance: (1) they function as *rules of inference* which help provide explanations and predictions of items of fact; (2) they offer *systematically unifying schema* for qualitatively diverse subject matters; and (3) they constitute means of *conceptual and symbolic representation* for the data of observational experience. These three functions are interdependent and are separated here only for analytical purpose.

(1) When considered as rules of inference, theories enable us to draw conclusions by their means. The purposes served by such drawing of conclusions are broadly of two types. First, the conclusions may refer to matters for which an *explanation* is sought. To possess an explanation is equivalent to exhibiting the way in which the item to be explained is logically derivable from other accepted or recognized states of affairs in a manner which satisfies the rules of the theory. Secondly, the conclusion of an inference as determined by theory may also serve as a *prediction* of what, if our reasoning has been effective, we expect to find in fact. In such cases predictions are a guide to action if confidence is already established in the theory. Where such confidence may not be sufficiently established or is lacking, predictions serve as possible *tests* for the theory itself. Confirmed predictions are explained in turn by the very process which showed them to be logically derived conclusions of a pattern of inference satisfying the rules of the theory. Statements expressing conclusions may be of the type which either describe individual occurrences or summarize, as generalizations, repeatedly observed conjunctions of traits. To take a familiar example, Kepler's laws as empirical generalizations of the orbits of the planets in our solar system are derived from and so explained by Newton's (as they also may be by Einstein's) theory of mechanics. When further supplemented by data giving specific information about the position and momentum of a particular planet, say Mars, at a particular date, the theory enables a prediction (or explanation) to be had for the motion of that same body at an earlier or later date in its career.

One may describe in either of two ways how the rules of inference which in effect *are* the theory in one of its roles, function for explanatory and predictive purposes. One may say that they are the fundamental premises *from which* inferences are made to the facts to be explained or predicted; or we may say that they are guides *in accordance with which* inferences are made to the facts to be explained or predicted. In either case it is important to note that they are insufficient as premisses or as guides when taken alone to yield the facts as stated in the conclusion. One needs, in addition, an instantial premiss or a statement of initial conditions, which, taken in conjunction with the theory, will yield the conclusion in question. A theory however, is not something of which it is appropriate to ask whether it is true or false. It is not, logically speaking, anything which can be considered to be a statement of fact or a series of such statements of fact. A theory rather is a pattern of ideas which *may be used* in interpreting the "givens" of experience. This pattern of ideas may be stated in sentences, but these do not refer to anything in the world of experience; they do not correspond to anything we may find either by simple observation or even by refined and elaborate experiment. It follows from this that if we wish to continue to speak of theory as a "premiss" of an argument, it must nevertheless be distinguished from the universal premisses of the classic syllogism which, like so many so-called laws, are statements of empirical regularity descriptively formulated. For this reason it would probably be less confusing to regard a theory as the rules *in accordance with which* we draw inferences *from* given facts *to* other facts.[1]

Two further remarks need to be made in order to clarify the sense in which we may regard a theory as constituting a set of rules of inference. First, it is necessary to distinguish the sense in which the specific equations of the theory may be spoken of as rules of inference as contrasted with the rules of inference associated with the particular mathematical calculi or natural language systems employed in the statement of the equations and the commentary accompanying them. The latter rules, for example, those belonging to some system of geometry, Euclidean or other, the differential calculus, the tensor calculus, or the syntactic and semantic rules belonging to the English or other "natural" language, are systems of general rules which are brought to bear upon the manipulation and interpretation of those symbols which are taken from these language systems and made to function in the particular theory under considera-

---

1 Cf. Toulmin, S., *The Philosophy of Science* (London, 1953), 42.

tion. The sense in which, however, we may say that the theory itself as expressed through its equations and supplementary textual commentary provides "rules of inference" is of a more specific sort. It prescribes the particular way in which we are to connect the symbols or terms occurring in the statement of the theory. Newton's equation for the Second Law, for example, prescribes the particular way in which we are to combine the symbols representing force, mass and acceleration. The second point to be made concerns the different degrees of generality of the rules constituting the theory of some subject matter. The field equations of the general theory of relativity, for example, state rules of a most general kind. The application, specification, or solution of these equations as holding for a static, spherically symmetrical field in the vicinity of a gravitating point particle (Schwarzschild solution) gives us a particular set of rules of inference which may then be further applied to describe the gravitational field surrounding the sun, and used to explain and predict the motions of material particles, for example, the planets and light rays in such a field. It was indeed these specific rules of inference, together with the background of the general field equations, which led to the well-known tests of the general theory.

(2) The second important feature to be noted with respect to theories is the way in which they serve to unify systematically wide tracts of qualitatively diverse subject matters. The fundamental equations of a theory are in effect a symbolic matrix or schema. As such they do not contain any terms or statements which directly describe either individual objects or events or classes of such entities. Theories, therefore, as we have already remarkd, are not only not singular statements, they are not even "descriptive generalizations" summarizing recurrently observable features of some specific subject matter. The statement of a theory is that of an abstract conceptual framework. It lacks any indication of the specific subject matters to which it may be applied. The scope of such application, the success with which it can be used in any particular case and the range or extent of such cases altogether, is something which has to be discovered by a process of trial and error. The scope of a theory is thus not something which is predetermined or stated by the theory itself. It is always a matter of finding out through experience how felicitously and successfully one can use the symbolic means provided by a particular theory in dealing with a given type of phenomenon. This is a

matter of professional judgment which, in turn, rests not simply on a trained "intuitive" sense but primarily on the pragmatic usefulness of what is accomplished thereby. In any case, all such judgments of fittingness are never more than relative and are always open to the challenge provided by a candidate in the form of a more promising theory. Normally, too, the greater the range and variety of cases that can be handled by a theory, the greater its value. On the grounds of economy and simplification alone it would be preferred for its capacity to unify and enclose within its grasp a wide variety of materials. It is, however, primarily because it contributes to our understanding that it is so valued; for understanding means noting underlying sameness of pattern in apparent irreducible diversity. The intellectual power of a theory as great as Newton's rests on the fact that the same basic equations and conceptual apparatus which they embody are used to make intelligible such varied phenomena, among others, as coiled springs, transmission of sounds, spinning tops, flowing rivers, moving planets, or the diffusion of gases.

(3) We turn, finally, to an examination of the way in which theories provide distinctive resources for conceptual and symbolic representation. Whatever be the ways things have of behaving, we cannot know what they are until we say what they are; and we cannot say what they are without using some system of symbols or other, some language or method of representation. The empirical element in knowledge never comes to us "pure." It is the task of symbolism in general to express what, how, where, and when we find what we do. But such symbolism is of various types and occurs on several levels even when we confine our attention to cognitive experience in its scientific form.

Already on the level of what we call ordinary experience or common sense, identifications and discriminations among the traits, processes, and objects encountered, involves the use of a moderately elaborate conceptual scheme. The ability to identify and discriminate various types and degrees of color, temperature, shapes, sizes, sounds, and motion, as well as to classify the great variety of objects and processes to which these qualities are assigned, all of this man articulates in ordinary language in a manner which suffices to raise him above the brutes. The concepts so employed on the level of everyday language are used for the most part in referring to and describing such objects and their properties as are found in common experience. "People," "houses," "trees," "circular," "green," "sweet," "loud," "swift," "hard," and the like are used in statements either about individual objects or about groups and classes of such objects or processes. It is a mark of such concepts that they are fashioned in the first place to hold for entities encountered in experience. The ranges of their meaning are not vacuous. Exemplifications abound. Such concepts serve us well for many purposes even though much vagueness and imprecision may be connected with their usage. Satisfactory communication and ordinary understanding are secured by staying within the core of clear meaning. Difficulties of various sorts—puzzlements, philosophical and otherwise, failure of communication, practical ineptness—arise when we venture beyond the relatively clear core of meaning and try to use these concepts where their twilight zones of meaning, the fringe areas of relative unclarity begin to show up.

One purpose served by science, even on a fairly "elementary" or pre-theoretical level in the refinement of our discourse about ordinary objects is the introduction of operational procedures of a quantitative sort which reduce vagueness by employing instruments and associated numerical terms for the ordering of our experience. The use of the ordinary thermometer, calibrated for distinguishing more accurately and objectively degrees of temperature than is the case where reliance is made on the feelings and response of the human body, is a familiar example. The enlargement of the area of clear meaning in ordinary concepts through advanced schemes of classification or through measurement constitutes one side of the scientific explication of ordinary concepts. Copper, milk, wood, are not simply what they are "known" to be for crude everyday experience; they are now in addition identified in more refined ways because of the efforts of the chemist, for example, who includes such data as melting points, or butter-fat content, or tensile strengths and the like.

The formulation and testing, too, of empirical generalizations or "laws" as these are commonly understood, becomes the central concern of science in its "descriptive" phase. Such generalizations are established through the use of deliberately contrived experimental procedures and increasingly refined techniques of quantitative measurement and statistical analysis. The point at which the distinctive conceptual resources of theory enter into this stream of knowledge, however, is not one that can be specified with exactitude. Yet there can be no doubt that what theory makes available, particularly on its more advanced levels, cannot be found either in common sense or in the early descriptive stages of science. This may be seen by considering the way in which an object of "ordinary experience" that comes under scientific scrutiny is, in our conception of it, actually encased in several layers of interpretation. What we call "the Earth," for example, is accessible to observational inspection in a variety of ways. The traveler, geographer, geologist, geophysicist and astronomer, among others, provide us with a mass of observational detail as well as with the conceptual tools for their interpretation. We say it is, for example, mountainous over part of its surface, that it is roughly spherical in shape, that it has a diameter of approximately 8000 miles, that it moves around the sun, that it possesses its own gravitational field, that it is subject to the gravitational attraction of the sun, and that it follows as a free body a geodesic path in space-time. While some of its "properties" may be said to be determined by inspection or measurement, others clearly involve the appeal to theory on an advanced or even controversial level.

In some cases, as in the above example, the concepts associated with theory provide means of handling objects or processes already identified in ordinary experience; in other cases, as in the "gene," "electron," or "expanding space," the concepts introduced by theory become significant only for experience and experiments achieved on a sophisticated level of observational experience accompanying the use of refined instrumentation. In any case, what we include under the heading of "theoretical concepts" are not, as we shall see later, constructs out of empirical materials in any simple or direct way.

What frequently is referred to as a matter of "fact" is, moreover, upon analysis (particularly in those areas where science has reached the level of competence in theory) discovered to involve a considerable amount of theory. The distinction in such cases between "fact" and "theory" turns out to be at best a relative one. What we call "observational facts" are those items of information which are relatively free of controversy with respect to what they summarize or represent. They are "theory-neutral." They are the data which any theory or diverse theories accept as unquestioned. It is the task of theory to interpret at least some of the "facts" as thus accepted in its own distinctive terms. The distinction is best seen when put into the context of inquiry. What we call a fact is that which is already expressed with the aid of some conceptual means, and particularly on an advanced level of science, with the aid of theoretical concepts. What we call the theory relative to such facts is the status of some further, more inclusive symbolic system which is available for additional interpretation of some of these facts. In principle, the privilege always exists to question the aptness of even the terms used in the "description" of such facts. What may have been unquestioned, can, if sufficient grounds exist for doubt, be re-examined to see if the very terms used in the report of observational data do not contain even indirectly certain theoretical constructs and interpretations with which it might be wise to dispense. A supposedly purely phenomenological or descriptive generalization like Boyle's law, summarizing the connections between the pressure and volume of a gas, contains implicitly a host of hidden theoretical assumptions and constructs even apart from the interpretation which may be given to the law itself by a theory such as the dynamical theory of gases. As Duhem has pointed out in a pregnant analysis of this law, the very descriptions of the measured values of the volume and pressure of the gas involve a variety of theoretical borrowings from areas such as optics, mechanics, and the like. Consider, by way of example, the recording of the volume of a gas under specified conditions and what it involves. "In order to form the first of these abstractions, the value of the volume of the enclosed gas, and to make it correspond with the observed fact, namely, the mercury becoming level with a certain line-mark, it was necessary to calibrate the tube, that is to say, to appeal not only to the abstract ideas of arithmetic and geometry and the abstract principles on which they rest, but also to the abstract ideas of mass and to the hypotheses of general mechanics as well as of celestial mechanics which justify the use of the balance for the comparison of masses; it was necessary to know the specific weight of mercury at the temperature when the calibration was made, and for that its specific weight at O° had to be known, which cannot be done without invoking the laws of hydrostatics; to know the law of the expansion of mercury, which is determined by means of an apparatus where a lens is used, certain laws of optics are assumed; so that the knowledge of a good many chapters of physics necessarily precedes the formation of that abstract idea, the volume occupied by a certain gas."[2]

2. Types of Theoretical Concepts

Theoretical concepts are not all of one type, and it will be useful for our present purpose to distinguish three groups. (1) In one group we find those theoretical concepts or constructs which are *schematized idealizations* of objects, properties, processes, or relations found in ordinary or direct experience. In order effectively to handle a complicated situation the appeal to theory will not only involve excluding those features of the phenomenon which on the basis of the given theory are considered irrelevant, but will require that a special technique of analysis and representation be used with respect to those aspects that *are* included. This technique consists in using concepts which, while suggested by experiential objects, can *never* be exemplified *in* experience. For they are in the nature of *ideal limits*. Such is the case, for example, in the way in which a planet is treated from a kinematic point of view as a *point particle*, or, as in Newtonian dynamics, as a *point particle endowed with mass*. One does not, indeed cannot find in experience point particles whether possessing mass or not, simply because what is *a point*, as a wholly non-extended entity, is an idealized tool of representation and analysis, *not* a discoverable object. The same holds for other notions of a similar theoretic status, like absolutely rigid bodies, perfectly straight line paths, frictionless engines, instantaneous velocities, perfectly isolated or insulated systems, ideal gases and so on. The great value of such concepts lies in their usefulness for purposes of analysis. The "return to reality," that is, to the actually observed or observable cases of experience is accomplished, among other things, by correcting or supplementing the theoretic account, by inclusion of the gross, non-ideal, qualitatively identified or measured traits belonging to such real occurrences. (2) Another group of theoretical concepts are those having primarily a mathematical significance. Such constructs need not be thought of as "idealizations" of objects or traits encountered in experience. Where, indeed, even these may in some cases be shown to be connected by way of origin in some indirect way with such materials, they do not depend on such reference for their tech-

[2] *The Aim and Structure of Physical Theory* (Princeton, 1954) 146; cf. Watson, W. H., *On Understanding Physics* (Cambridge, 1938) 62.

nical meanings. Such theoretic concepts may be illustrated by the appeal to "tensor fields," "radius of an infinite sphere," "world-lines in space-time," "curved (or 'expanding') space," and the like. These function in the context of physical theory as instruments of calculation and representation. Very often names will be given to such constructs which recall some analogue in a domain already familiar through experience of a qualitative or measurable kind or even with some other theory's concepts. Such is the case when we speak of "expanding space" and think of visual analogues in "expanding balloons," or the way in which "the potentials of a gravitational field" or "the velocity of a point particle" in Einsteinian theory are "generalized" from the Newtonian usages of these concepts. There is no doubt value both for creative thinking and pedagogic purposes in preserving a sense of such "continuities" of thought. But also there is a danger. For the propensity to expect of the extended or new theoretic constructs *all* those features found in the "primary" cases of their usage, proves to be a source of confusion, particularly noticeable in popularizing accounts of these matters. The only remedy for such confusion is the painstaking examination of the precise ways in which these concepts are defined and a careful regard for the exact ways in which they function in such theories. Where a correlation or "dictionary" (as Campbell calls it[3]) is established between theoretic concepts and terms having an independent empirical meaning (though frequently designated by the same words) it is important, nevertheless, to keep the two distinct, and to recognize the difference in source and sanction which applies to each. In the dynamical theory of gases, for example, "pressure" is defined for the theoretical concept of gas-particles in interaction with one another and with the walls of a container; to correlate such "pressure" with that which is also called "pressure," as something measured in a laboratory by some instrument and as applied to an actual gas, does not mean that the two can, in fact, be identified. To take another example: the definition in theoretical cosmology of the "distance" of a light source from an observer, which involves making use of the notion of a scale-factor or radius of curvature of space (as is done in relativistic approaches) is to be kept distinct from, even though it may be "co-ordinated" with, the astronomers'

[3] Campbell, N. R., *Physics; The Elements* (Cambridge, 1920) 122.

operational definition of distance as defined in terms of apparent and absolute (photometric) magnitudes. In any case, it is not necessary that *all* theoretic concepts receive a "dictionary" co-ordination with concepts already possessed of a direct empirical or operational meaning. Many, indeed, have their value *exhausted* by the role they perform intra-systematically by guiding calculations and symbolic transformations and which, as such, have *no* counterparts or correlates in the subject matter to which the theory as a whole may be applied. (3) We come, finally, to a group of theoretic constructs sometimes called "scientific objects" or "inferred entities." Included here are such concepts as "electron," "absolute lumeniferous ether," "gene," and the like. There are two different ways in which these may be regarded, depending on the possible answers that may be given to the question "Do these objects or entities exist?" By existence we may understand the possession of dispositional properties, of capacities for independent interaction in ways appropriate to the entity involved with other entities, including the possibility of disclosure of such existence to sentient and inquiring creatures such as ourselves. It is possible to regard the introduction on the level of theory of concepts referring to such "entities" simply as explanatory devices, useful in the formulation of rules of inference but without any correlates in the world of independently existent objects and processes. In such cases these "entities" may be grouped along with the theoretical concepts of the second of the above-mentioned groups, since they are not in any way basically different from them. It may be, however, that in addition to regarding such theoretic entities as explanatory devices one thinks of them as terms which may be coordinated to or predictive of independently existent objects manifesting themselves to us in various experiential ways. The clicks on a geiger counter and the visible tracks in a Wilson cloud chamber are then not simply part of the data to be interpreted with the aid of theory, but signs for the presence of specific types of objects and occurrences disclosing themselves to us by these evidences. The confirmation of the existence of such objects or processes may be thought of as resulting from the predictive powers of theory. Even in such cases,, it would be more accurate, despite the common practice of using identical terms to designate such existent entities and the theoretic constructs serving a purely inferential and symbolic role, to keep the two quite distinct from a logical point of view. In the absence, moreover, of any confirming evidence as to the existence of such entities, their theoretic value need in no way be impaired. For on the level of theory the determination of whether use of constructs like electrons, phlogiston, the absolute ether, and the like, are to be judged successful or not is one which is decided not by reference to these concepts taken in isolation, but in terms of the applicability of the theory as a whole of which they form a part.

3. Types of Theoretical Statements

What is called "theory" in a broad sense actually includes various types of statements or groups of statements, of which for present purposes we single out for discussion *principles, laws,* and *models.*

By a *principle* we shall understand that type of theoretical statement which expresses a basic conceptual orientation or technique of representation upon which other statements such as laws and models belonging to the same theory in an inclusive sense are dependent, in part, for their meaning. The statement of the principles of a theory is closely connected with the role, mentioned earlier, which theories perform as a symbolic matrix or schema capable of being applied to qualitatively diverse subject matters. Principles are not always called by that name and in order to identify those statements which in fact serve as such, one has to determine whether in the theory concerned they are basic or not. In Newtonian mechanics, for example, the law of the pendulum and the law of falling bodies are not basic in the present sense. They depend for their very meaning, in part, upon the three so-called Laws of Motion and the Universal Law of Gravitation, whereas the converse is not true. It is these latter statements which properly should be designated as the Principles of the System. Similarly, the Principle of the Rectilinear Propagation of Light is fundamental for the entire discipline of geometrical optics, the Principles of Equivalence and Covariance underlie the whole structure of the general theory of relativity, and so on.

The term *"law"* in science has been used in a great many different senses. Sometimes it is used to designate that type of statement in which there are no explicit occurrences of theoretical terms and which formulates an empirically discovered regularity, a "descriptive generalization." In other cases we speak of a "law" as a statement which expresses a specific type of regularity, one indeed which may be suggested by recurrent observation or controlled experiment, but whose formulation involves an explicit use of concepts of a theoretical sort and therefore depends upon the appeal to some principle of the theory concerned. It is this latter sense of "law" which is here intended in speaking of it as a part of theory in an inclusive sense. Indeed, what may be simply an empirical regularity containing exclusively descriptive terms at one stage of inquiry, may, with the advances accomplished through the systematization introduced by theory, become re-expressed in the symbolism of such theory and incorporated now within it. Such is what happened, for example, to Archimedes' law of the lever or Galileo's law of falling bodies when incorporated within the system of Newtonian mechanics, or Boyle's law and Charles' law when incorporated within the framework of the dynamical theory of gases, or Faraday's law of induction when made part of an electro-magnetic field theory.

With respect to "model," once more there are a variety of usages connected with the term. As used here, we shall think of a model as that form of specification of a theory which is so constructed that it will be used in the account of some specific and restricted range of identified subject matter. Just as laws taken singly fill out and restrict the range of applicability of principles, so, in turn, one may think of models as further restrictive specifications of one or more laws as applied to the conditions exhibited in a specific type or collection of objects. In this sense, one may speak of a model of the solar system, of pulsating stars, or of the universe itself considered as a physical system. There is no requirement that models be constructed exclusively of theoretic concepts of one type. In this respect, the insistence of many nineteenth-century physicists that all models be mechanical ones, that is, employ the theoretic concepts of Newtonian mechanics for all physical phenomena, was a hampering and unsuccessful restriction upon the conceptual materials to be used. Also, of course, the even more stringent demand that models be "visualizable," violates the conditions under which theoretic concepts are to be employed. Such a demand is at once irrelevant, and indeed, most often unrealizable. Failure to satisfy it does not in any way impair or nullify the intelligibility and possible usefulness of such models.

4. Critique of the Classical Philosophy of Science

The classic philosophy of science, developed in ancient Greece and persisting in modern times, borrowed its essentials from Plato and found a comprehensive and authoritative statement in Aristotle. It continued as the dominant view well into the nineteenth century, despite its alleged overthrow by the founders of modern science in the seventeenth century. This philosophy is founded on the fundamental belief that Nature has a fixed, determined, intelligible structure; that the task of science is to discover this structure and to give it articulation in a deductive system and thereby provide explanations for specific details and occurrences encountered in experience.

For Aristotle, reality was thought of as consisting of substances whose forms are the common intelligible essences of individual things. There are multiple, uniquely intertwined natural classes or kinds to which individuals belong and in terms of which their properties may be understood. Explanation for science is to be accomplished through a process of demonstration. For Aristotle this meant using our powers of intuitive reason in what he called "induction," to discern the intelligible essence of things. Such essences are to be stated in the form of definitions and made to serve as axiomatic principles for a deductively systematized body of knowledge. The axioms or definitions serve as the ground for necessary inferences to the implicated properties of the subjects studied. The goal of such an explanatory scheme as a demonstrative system was best seen in geometry. The same basic pattern, however, is one to be followed in setting out knowledge acquired in all domains of natural fact. On the primarily classificatory level of Aristotle's biologically oriented view of science, the syllogism was the primary framework of proof for the properties belonging to the various species of things.

The conception of science worked out in modern times, from the late medieval period on, culminated in the great achievements of the seventeenth century and continued to guide and interpret the work of science in various fields up to the end of the nineteenth century. It was a conception of science arrived at by making certain changes in, while retaining other important features of, the classical view. Retained from the classical view are two essential ideas: that explanation is achieved by exhibiting the logical deducibility of a fact from well-established premisses, and that there is a unique, specific, determinable, and intelligible structure in Nature with which science is concerned and which it is competent to articulate. Despite the emphasis on experiment and mathematics in modern science, there is continuity with the classic tradition in adherence to what Dewey has called a spectator view of knowledge; man's reason is like a great eye, gifted with understanding and directed to giving us a veridical total picture of the world spread out before us. We must strive, it is thought, to obtain that one view of Nature which will truthfully express its essential structure. The crucial changes, effected in the classical picture, revolve about the method appealed to for warranting it. Instead of the search for essences inherent in the class-structure of substances, modern science directed its attention to the discovery of laws or regularities expressive of the *interconnections* and relations among the *measurable* factors of phenomena. And instead of reliance upon a process of immediate intellectual intuition to sanction the adoption of the logically certain, axiomatic definitions of essences, modern science turned increasingly to controlled observation and experiment. The appeal to experience is at once for the purpose of yielding the materials for inductive *generalizations* (rather than definitions) that in turn would serve as the warranted premises for deductive inference, and as the basis of choice from among competing hypotheses as to which proposed generalization might indeed state the true "law."

Within this broad framework of ideas the differences between "rationalists" and "empiricists" were family differences, each school stressing one or another aspect of the entire enterprise. For the rationalists, the emphasis is placed upon mathematics as the key with which to unlock the secrets of Nature. As hailed in the advent of mathematical physics, the way seemed open for abandoning the gap Plato insisted on making between the world of physical reality as one in which only opinion could be possible, on the one side, and that of pure, intelligible Form, on the other, about which demonstrable and infallible knowledge could be had. Now at last one could aspire to explore with all the certainty of mathematical science the world of physical fact. The two domains coalesced and the rationalism of the seventeenth century philosophers, so clearly expressed in the goal envisaged by Descartes and Leibnitz of a universal mathematics became a widely shared view. Mathematics is uniquely suited to summarize and express the fruits of measured, quantified data which modern science stresses as its own basis of advance beyond the classificatory, qualitative stage of science in the ancient world. Mathematics for the rationalists, however, is not simply a convenient device of calculation or symbolic technique, an *ens rationis*. It is rather the most powerful means for directly expressing the *ens reale,* the objective intelligible structure inherent in Nature. Galileo, who is typical in this respect, for all his antagonism to the Aristotelian doctrine of essential natures, accords to the patterns studied by mathematical physics the same ontologic status and sanction which essences had in the earlier philosophy.

Meanwhile, empiricism, particularly in the British tradition, in its effort at doing justice to the appeal to experience as a basis for warranting judgment and as the source of inductively generalized laws, struggles, though never successfully, to adjust its insights into what such an appeal would seem to entail with the rationalist presuppositions that it inherits with the classic philosophy and that it is unwilling and unable to shake off completely. Thus, on the one hand, one finds in British empiricism the demand at work for a certain, unique warrant for judgments. This takes the characteristic form of a search for the reduction of "complex ideas" to, or their derivation from, intuitively clear and unexceptionable sense-data, instead of, as with the rationalists, a search for intellectually apprehended "clear and distinct ideas." On the other hand, there is the evident limitation in any appeal to observation and experiment: no appeal to experience as a sampling basis for generalization or as a testing basis for elimination or retention of hypotheses can ever be expected to yield certainty. The adoption of any formulation of what the regularities of nature are, is always subject to modification because of the limited range of human experience. All empirical statements which purport to generalize or make predictions about the future can only be probable. Here, for example, in the clash between the classic demand for certainty and the limited assurances of the appeal to experience is born Hume's problem, the problem of "justifying induction." It is a problem to be "solved" only by abandoning and stepping outside the framework of presuppositions derived from the classic philosophy and turning to other conceptions altogether.

The conception of science which has gradually been coming into focus since the end of the nineteenth century (although there are roots for this movement as far back as Kant, at least) is one which has gradually served to weaken the grip of the classic, rationalistically oriented, philosophy of science. It is a conception of science to which various schools have made their contribution—positivism, conventionalism, pragmatism, operationalism, and current analytical studies of language. Contributing too, to the breakdown of the older tradition were certain revolutions in science itself in the nineteenth century and the early part of this one, particularly the rise of the non-Euclidean geometries, and in physics, the overthrow of once-confident mechanical theories. Such developments, and those which followed in their wake, have enforced the lesson of the necessity for recognizing the existence of a multiplicity of available possible "ordering" devices and the spuriousness of any claim to self-evidence that can be made in behalf of any one scheme. Close study, too, of experimental procedures has brought with it recognition of the inherently fallibilistic character of all generalizing empirical judgments due to the uneliminable tentativeness and incompleteness of all appeals to observation. And increased awareness has come for the presence of conventions in science arising neither from reason nor sense yet playing a significant role in the interpretation of empirical data and in the very formulation and use of theories.

The classic philosophy must, accordingly, on the whole be judged inadequate because of its failure to do justice to the constructive nature of the concepts and the conventional factors belonging to scientific theories. Regard for the status of theories as so many languages or symbolic systems enforces the conclusion that there is no warrant for assuming that things themselves possess a structure in the sense that this corresponds to or is identical with that which is expressed by some unique mode of symbolic expression. At best there are as many ways of expressing what the facts are in theoretic terms as man's ingenuity allows him to devise. It makes sense to say that some are better or worse than others, but it makes no sense to say that there is an absolutely best one. Theories are apt or fitting but they are not as such *true*, where truth is taken to mean "correspondence" of symbol and existence. Indeed, to speak of fittingness can itself be a misleading analogy. In the case of a suit that is made to fit a man, we can measure and describe the body of the man *independently* of the suit. But in the case of theories, as in the case of languages, it is meaningless to think of nature as possessing its own code which might in principle be explored independently of our symbolism and a test carried out to see which one of our human devices most accurately "matches" the "real" one. Taking seriously the symbolic and constructive character of the theories of science means re-orienting our conception of truth as a goal for science. It means giving up the "spectator" conception of knowledge and the "one-shot" criterion of adequacy.

But, it will be asked, does this not in effect remove all incentive and purpose from science? If there is no absolute truth to be discovered, how can we measure progress in obtaining knowledge? What is there to justify dissatisfaction with what we already have, or inspire search for the new? In what way can the new be *better?* These questions may be answered, at least in part, by considering what are in fact the operative criteria of evaluation for theories. Here we are helped by reverting to our analysis of the roles which theories perform as rules of inference, systematically unifying schema, and conceptual tools of representation. Theories are valued for their *predictive power,* their *degree of comprehensiveness,* their *economy of conceptual means.* It is these qualities which take the place of "truth," at least as far as theories are concerned. They serve as regulative standards of evaluation without any presumption that there is or might be any one theory which might satisfy these criteria in "the highest degree." Indeed, strictly speaking, there is no meaning to the phrase "highest degree" in this context: there can only be judgments made of *comparative* worth among *actually available* theories offering varying satisfaction of these criteria.

The *predictive power* of a theory is another way of expressing its applicability. The rules of inference constituting a theory are conventional; as such they cannot be proved or disproved; they cannot be shown to be either true or false. Such terms of evaluation are irrelevant to them. One can only ask: "If we accept certain statements of matters of fact as true (if the 'initial conditions' of some situation are truthfully reported) and we *use* the rules of inference provided by a given theory in drawing the consequences of those conditions and matters of fact, then do we as a matter of experienced fact find what the inference leads us to expect?" Is the conclusion derived *by* the theory *from* the data such that it can be confirmed *in* experience? If it is, then we can say that the theory has proved its *usefulness* in these or similar circumstances, not that we have verified the theory. Rules of inference are devices of thought whereby we undertake to link facts, without themselves being facts. Success in handling some subject matter by means of a particular theory does not mean possession of some structure "inherently" by the subject matter corresponding to the theory; all we can say is that the subject matter is such that we can reason successfully about it by using the given theory; it does not exclude the possibility of other theories, employing altogether different concepts and rules being equally or perhaps more successful. Tests, therefore, of truth as developed in traditional epistemologies are irrelevant to theories. One does not look either for their "self-evidence" or for their own direct empirical corroboration. Nor can one deduce the rules from anything more fundamental. At best we may undertake to supplement the rules of one theory with those of another. Once more the question of the validity of such a combination can only be tested pragmatically, not simply by any internal examination of consistency. When use is made, for example, of the conjunction of rules "A" and "B" to infer or predict fact "y" when given fact "x," do we indeed find "y?" If the answer is no, then assuming that the derivation or calculation is correct, the *joint* applicability of "A" and "B" is to be doubted. Either "A" alone or "B" alone may continue to be useful in predicting other facts, or the conjunction of "A" with "C" or "B" with "C," or some altogether different set of rules may now have to be invoked to explain "y." Such indeed was the situation classically illustrated in the Michelson-Morely experiment. What Einstein realized was the essentially *incompatible* combination of the velocity-addition theorem from the theory of Newtonian mechanics with the doctrine of an absolute ether as contained in Maxwell's theory. The adoption as a fundamental rule of the Special Theory of Relativity of the constancy of the velocity of light for all observers in uniform relative motion, was now one of the means to be successfully employed in explaining the negative outcome of the Michelson-Morely experiment.

Another criterion for evaluating theories is that of *degree of comprehensiveness*. We shall say that one theory is more comprehensive than another when it can handle, i.e., successfully predict and explain, all of the facts which the latter can, and, in addition, it can successfully deal with other facts that do not fall within the range of the "narrower" theory. The more comprehensive theory may also be spoken of as a richer theory, meaning not simply that its conceptual resources involve more extensive and subtle differentiations than some other, but rather that, by these means, it is able to encompass a wider range of instances and types of subject matter within its grasp. One may say that the more comprehensive theory "extends over" the less comprehensive one.[4] It is in this sense that electromagnetic theory is more comprehensive than electrostatics, physical optics than geometrical optics, general relativity than special relativity. It is in the same sense, too, that relativity theory is more comprehensive than classical mechanics. Newton's theory is applicable only to a first approximation; Einstein's account "reduces" to the Newtonian one under certain special conditions. It makes provision, however, for a far more refined account of dynamical facts. Its use of ten potentials rather than one, enables the later theory not only effectively to handle all cases of dynamical phenomena that the earlier one had, but many others as well that either were not explained at all or, in some cases, even recognized to exist.

By *economy of conceptual means* as a desirable feature of theories is meant logical simplicity. This is exhibited in the ability to unify within a single formal scheme containing relatively fewer theoretic concepts and without the introduction of *ad hoc* assumptions the same or a larger range of facts than is possible by means of other theories. Once more an example is at hand in the case of relativity theory. As contrasted with the situation in classical physics which employs the concepts of both inertial and gravitational mass, relativity appeals to but a single notion; or again as contrasted with the classic use of the idea of gravitational forces in accounting for the motion of a body in a gravitational field, general relativity abandons the concept of such forces altogether and in its place employs simply the geometric concept of curvature of space-time as determining the geodesic paths of such bodies. Metric coefficients holding for the "geometry" of a space-time region become identical with the gravitational potentials. At the same time, within the general theory of relativity the same type of geometrizing treatment is *not* available for electrodynamics of the motions of charged particles in an electromagnetic field as it is for mass particles in a gravitational

---

[4] Cf. Törnebohm, H., *A Logical Analysis of the Theory of Relativity*, (Almqvist and Wiksell, Stockholm, 1952), p. 205.

field. The goal of a unified field theory as pursued by Einstein and others would accomplish the type of logical unification which is characterized by appealing to the same basic conceptual means for both types of subject matter which is now in use for just one.

Chapter 5

The Observable Universe and Cosmological Models

1. Meanings of "the Universe"

THE OBSERVABLE materials made available at a given time by astronomical research on its most comprehensive scale and providing thereby the data or subject matter for incorporation within a complete cosmological account has to do with what is frequently referred to as "the observable universe." What, therefore, is called "the observable universe" at one stage of astronomical knowledge can and does show important differences from what the *same* phrase designates at other stages of its development. At the present time, for example, the whole mass of information obtained through the use of telescopes and auxiliary instruments like the spectroscope, about the population of the galaxies, their spatial distribution, average masses, spectra, luminosities, shapes, stellar and material composition, and so on, all this refers to what is designated as "the observable universe." Since, however, it is not in any way presumed that present instrumental resources have exhausted all there is to be observed, that the limits of present observability which are set by the degrees of refinement and power of available instruments, are at all final and absolute, there is not only the incentive to develop more sensitive instrumentation, and thereby extend the horizon of what is actually observable, there is also meanwhile the challenge, on the basis of what *is* known, to anticipate and predict what may be disclosed when such resources do become available. Moreover, the materials yielded by observation pose a variety of problems which call for interpretation The interconnections among the observable facts, the explanations for the specific traits found of a qualitative and quantitative kind have to be sought for and provided by some conceptual scheme. To offer the basis of extrapolation beyond the available data and to give the conceptual tools of interpretation for the materials of observation is the task of theory. In cosmology this double role is fulfilled by the construction of "models of the universe." Such "models of the universe," sometimes referred to as "universes" for short, include as prominent examples in current discussion "the expanding universe model(s) of general relativity" (Einstein, de Sitter, Lemaître, Eddington, McVittie, etc.), "the steady-state model of the expanding universe" (Bondi and Gold, Hoyle), "the model of the universe based on kinematic relativity" (Milne) and others.

The data obtained from the domain already explored, "the observable universe," is sometimes referred to as a "sample of the universe." [1] To be acceptable, however, such an expression has to be understood in at least two different ways which point in opposite directions and which should not be confused with one another. On the one hand, we may wish to assert simply that it is a reasonable procedure to use the data accumulated about, for example, the galaxies already observed as a basis for predicting the properties (spatial distribution, masses, spectra, luminosities, and the like) of the galaxies which we may expect to discover as our present instruments probe more deeply or as more powerful ones which may be built provide us with fresh and enlarged data. In this sense the region of space or collection of galaxies at present observable is used as a sample for predicting what one expects to find for a continuingly larger collection of galaxies that is also in principle observable. From this point of view the "whole" universe of which the region at present observed is a sample is not a "completed" whole or an "intelligible" whole, but a name for an *ever-increasing observable* domain or collection of objects. On the other hand, another role which the evidence provided by the "sample" may perform is to be seen in connection with the use of *models* of the universe. Here the function of the accumulated data is not to serve as an empirical basis for predictive inference to a *wider collection of similar objects* but instead to serve as a basis of selection from among several open theoretic possibilities *of interpretation*. Where, for example, some choice in the selection of a model is available

[1] Cf. Hubble, E., *The Realm of Nebulae*, 84; *The Observational Approach to Cosmology* (Oxford, 1937) 1, 18.

in terms of the density of the "fluid" or material taken as "filling" the model, the observed value of the density, if this is thought to be sufficiently well established, may serve as a guide in choosing the appropriate model. In selecting these and other specific values for the *theoretic* model, it is hoped not only to be able to provide a consistent and unified interpretation of the available data but to be able to guide successful predictions for hitherto unobserved phenomena as well. What is denominated "the whole universe" on the level of theory is *not*, however, something which either *is* or *ever could be* empirically observed. For what the term "the whole universe" means is what is signified by the entire set of devices of representation and mathematical or other symbols which are employed in a given theory by which it is proposed to interpret the data available concerning "the observable universe" and which are used in guiding predictive inferences concerning the regions as yet to-be-observed. The "universe-as-a-whole" thus does not constitute a name for some object or entity which exists antecedently to or independently of our inquiry and whose essence or structure we are trying to discover and articulate. Although cosmologists like many others very often talk as if there is such an object and that it is with it that they are concerned, an examination of their actual procedures discloses that what they in fact do is not in any way altered if one surrenders this belief or assumption. For what they in any case *do is to construct a symbolic scheme* which they employ with reference to the empirical data which they do possess. And an examination of what goes into the stipulation of the meanings assigned to the several conceptual parts of this symbolic scheme reveals that these do not in fact depend, either for the individual parts of the scheme taken singly, or their combined use as a systematic unity, upon the assumption that there is some object to which they refer and with which they can be made to correspond. Identification of "the whole universe" provided by theory is one accomplished by *definition* not by observation. It is not, therefore, a domain open to increasingly more adequate *inspection*. It is a conceptually "completed" whole. What distinguishes one theory from another is the way in which the "whole" is defined. The role of observation of the "sample" is to offer an empirical basis for determining the applicability of such schemes, for testing the adequacy of the variety of models entertained.

The distinction between the senses in which the universe is thought of as something which is in part or in principle open to observational inspection as contrasted with that which is conceptually defined by means of theoretical devices, and where the latter are to be used as inference guides for the former, may be further appealed to in helping to resolve a question which has been raised in recent discussions, as to how "the universe" is to be *defined*. Bondi raises the question as to what is meant by "the universe" as follows: [2] "Accepting the postulate that the velocity of light is the maximum velocity at which influences can be propagated, then it is evident that all events on or within our past light cone may affect us, and they certainly must be included within our definition of the universe. If we extend this to include all events which have been or will be affected by us, then we arrive at the largest set of events that can be considered to be physically linked to us. It is this set that is usually considered to constitute the universe. Some authors, however, consider a different set, viz., the largest set to which our physical laws (extrapolated in some manner or other) can be applied. This set may include points that have never been and will never be physically linked to us and may exclude points which are so linked. The physical significance of this 'universe' is therefore not very clear, but it should be remembered that even with the definition adopted above the actually possible observations concern only a very limited part of the universe."

The answer in general from the point of view of logic that one might give to the question just raised does not consist in attempting on *a priori* grounds to choose one or another of the possibilities indicated. For the question has to do primarily not with the universe as something actually or in principle observable, but with the models or theories to be adopted for purposes of representation and inference. And for such purposes *either* of the modes of representing the universe indicated above is a *possible* way of doing so. There is nothing sacrosanct or inevitably correct about a theory which imposes the velocity of light as an upper limit to all "transmissions" of physical influences. It is as much of a theory as one which envisages the possibility, on grounds distinctive to itself, of referring to events and objects which lie outside the light cone or beyond the range of physical "influence" of, or communication with, our own immediate physical environment. Which conception we shall adopt of what constitutes "the largest set of objects" that we are to identify as "the universe" is ultimately a question of which theory *as a whole* proves more successful. Neither one has a premium on how the universe *must* be defined, or represented. And even if a particular theory should see fit to introduce the notion of objects lying beyond the limiting horizon of what is set by the velocity of light, such a theory might prove fruitful in terms of the specific predictions it does make that *are* confirmable. While, therefore, a theory may at some stages of the elaboration of its ideas require reference to what cannot in principle be empirically detected, such a situation need not in fact be fatal to the theory in question. It is not necessary to demand that *all* the concepts and representational devices introduced by a theory be given an empirical reference. It is sufficient if the theory is able in virtue of its total conceptual apparatus in conjunction with the facts that *are* observable to lead successfully to the prediction of *other* observable facts.

The relative halting place of cosmological inquiry is marked by the possession and use of a unified, coherent, and empirically successful model of the universe in application to available, relevant observational data. In such a case one may speak of the outcome of such inquiry as providing a knowledge of "the universe." This phrase, however, if it is meant to identify an *object* of some kind, must be understood in a special way. As an "object," what is called "the universe" is a short-hand expression for all those distinctions of sense and theory, the system of connections, facts, and usages which refined and tested thought has come to accept in this domain. It means nothing apart from this funded background. It is the product of what men come to know when their observational experience is articulated and guided by a successful system of thought. It is an object in the sense in which Dewey has pointed out we speak logically of "that set of connected distinctions or characteristics which emerges as a definite constituent of a resolved situation and is confirmed in the continuity of inquiry." [3] "The universe" is thus in the same way the concern of the cosmologist as "life" is for the biologist, "mind" for the pyschologist or "matter" for the physicist. These are "objects" only in a highly telescoped

[3] Dewey, John, *Logic* (New York, 1938) 520.

sense. These various disciplines have as their respective subject matters, data which they seek to organize and explain. The outcome is an account which may be summed up by saying "Life (or matter, or mind) is such and such." It is the conjunctive assertion of these statements of "such and such" which constitutes what is known as and meant by "matter," "life" and "mind." And the same kind of thing holds for the way in which "the universe" is to be understood.

Before proceeding, it may be well to take note of a possible objection to the foregoing as a way of describing the enterprise of cosmology. It may be said that we have tended to slur over the distinction which must be made between the universe proper as it exists in its own right, on the one side, and, on the other, the variety of accounts, whether observationally descriptive, theoretical, or their combination, which may be given of it. After all, we are tempted to say, while surely there are many possible ways in which the universe can be or has been described, it is still about the same universe confronting us that we are talking. To use a simple analogy: one might say that though there may be variety, even disagreement in the accounts given of the properties belonging to John Smith, the disputants must at least be in agreement about the fact that it is with respect to John Smith that they are offering their respective accounts. So too with the universe. One must have some commonly agreed upon subject with which inquiry is concerned, even though there may not be any agreement in what is said about the subject. In the sentences, for example, "The universe is finite spatially" and "The universe is infinite spatially," assuming that the predicates are clearly and univocally defined, we should be able to assume that despite the presence of these contradictory predicates, it is about the same subject that these sentences are dealing. One should, so it would seem, be able either to make reference to or identify the subject "universe" even if all else is in question about its properties.

To this type of objection a number of replies may be offered. If by "the universe-as-it-exists-in-itself" is meant some cognitively inaccessible *ding-an-sich,* then such a metaphysical usage is wholly divorced from any meaning which can be assigned to a phrase like "the universe" which is guided by attention to the conditions of scientific discourse and inquiry. On the other hand, if by "the universe" is meant a subject which is cognitively accessible, then it needs to be pointed out that the identification of such a "subject" (where such language is used at all) is not possible without ascribing to "it" *some properties* in the very process of identification. Insofar as some meaning can be attached at all to the name "the universe" it is because some of its traits are agreed upon. This may be seen even more clearly when we put the whole question, as it is desirable to do, into the context of inquiry. What on the traditional Aristotelian logic is regarded as a *subject* which is thought of as having a status independent of and antecedent to all inquiry, is instead to be seen as a *subject matter* which marks the starting point for inquiry. As subject matter for inquiry, "the universe" is that which is identified by means of data which in part is constituted of the fund of accepted distinctions with respect to observed facts and conceptual meanings *to be used* in further inquiry, and in part is of an observational kind which being problematical and inviting explanation and extension of its range, poses the need for *further* inquiry. When seen in this light, we are prepared to note elements of both relative fixity and change in the subject matter which is, in this sense, "the universe" for inquiry. It *does* make sense to say that after all our various theoretical accounts are about "the same universe." For what is meant by "the same universe" is that which is identified by means of those commonly agreed to and recognized data with which any responsible theory has to deal. These are the facts of observation together with those items of conceptual formulation belonging either to common sense or refined science used in the interpretation of these observational facts which are not, at a given stage of inquiry, called into question. However, while there is some element of continuity or relative fixity in the materials entering into the identification of the universe as subject matter, there is also, of course, when we survey the whole stream of cosmological inquiry in its historical aspect, significant change. What belongs to the commonly accepted data today is different from what such data was in earlier epochs and is sure to be modified with ongoing research in the future. Scientific cosmology, as distinguished from purely speculative efforts at determining the nature of the universe, occupies itself with a task capable of progressive fulfillment, that of gathering observational data and looking for theoretical devices that will help in interpreting and augmenting the store of such data. In constantly

re-assessing its own efforts it makes progress in both directions, without nevertheless claiming an ultimate, unique end
to its search. The meaning or meanings of the phrase "the
universe" is best sought for and assigned to it in terms of
this many-sided activity, and the results of such activity at
different historical epochs.

2. Relativistic Cosmologies

We shall now undertake to illustrate in some greater detail
the manner of construction and operation of cosmological
models as these apply in the interpretation of the observational
materials of astronomy on its most comprehensive scale. For
this purpose we shall turn to some of the more general features
of the development of models of the universe that constitute
"relativistic cosmologies." No presumption is intended that
such models are the only successful ones, nor that the future
of the discipline is necessarily tied to their further multiplication or refinement. Nor will the attempt be made to explore
and assess the technical details with which investigators in the
field are concerned. Rather such interest as this summary
might have is one of giving an illustration of a method at
work.

Models of the universe that belong to the group of relativistic cosmologies have their common source in Einstein's
general theory of relativity. This theory is contained in a set
of field equations which provide its broadest schematic
framework. These equations, formulated in the language of
tensors, state a general relation between the four dimensional
space-time "geometry" of the field and the physical content
of space-time, namely, matter and energy, which, distributed
throughout the field, determine its metrical structure. The
importance of these equations is to be found in the fact that
they contribute to the goal of stating a pattern of relations
with respect to the entire realm of physical phenomena. In
particular, they permit the precise determination of those
aspects of a total field-physics which involve dynamical and
gravitational phenomena. Out of the original equations of
the general theory of relativity, as these apply on the level
of mechanics, two broad directions of application and specification may be distinguished. In one direction, that which
served as the first opportunity for experimental confirmation,
a rigorous solution was obtained to the field equations that
permits an application to the dynamics of a relatively re-

stricted field like that of the solar system. Here, in what is known as the Schwarzschild solution, particular values are given to the variables of the general field equations appropriate to the type of field we find in the solar system—a single massive "particle" (the sun) in the field of which the behavior of the remaining particles (material ones like the planets and those representing light) is examined. Out of this analysis came the understanding not only of those details of celestial mechanics already made intelligible by the Newtonian theory of gravitation, but a number of other facts as well. The famous experiments relating to the red-shift in the spectral lines of a light source due to the gravitational potential existing at the surface of the sun, the motion of the perihelion of Mercury, and the "bending" of a light-ray (or deflection of the path of light particles) in the vicinity of a strong gravitational field like that of the sun, were all regarded as confirming instances of the particular solution mentioned above and thus indirectly of the field equations as a whole.

The other direction of application of the field equations, of particular interest to us here, has been in the field of cosmology. The problem may be stated in a general way as consisting in the attempt to find a particular solution of the field equations which would serve as a model of the universe as a whole. Such a model consists essentially of two parts: a specification of the particular metric (line-element) to be introduced into the field equations and a determination of the values of the energy-momentum tensor appropriate to the metric field as thus defined, i.e., the special values of the matter and energy components in terms of which the distribution of matter and motion to be found in the observable region could be made intelligible. Among other things, the items tied together and systematically interrelated by such a theory include the behavior of material and light "particles" in the universe, the type of spatial curvature possessed by the universe as a whole, and its overall temporal features. A relativistic theory of the universe is comprised then of the following chief elements: it employs (1) the field equations of general relativity for which is specified (2) a particular form of line-element (or metric) which in turn is said to characterize (3) the material comprising the universe as a whole. The introduction of (2) and (3) into (1) yield a solution available for comparison with observed fact. Such a solution constitutes a model of the universe which in effect serves as a set of rules of inference providing explanations for observed fact or, when looked at in the context of inquiry, as an instrument for the prediction of various items of fact that are to be expected if the model is an apt one.

What now are the evidential grounds and methodological criteria for justifying the theory itself? It is in the kind of considerations introduced here that one discovers the touchstone for the philosophy of scientific method employed. Let us examine briefly some of the typical grounds of justification that are introduced in support of models constructed along classical relativistic lines.

Justification for appeal to the field equations of the general theory of relativity as a basis for "deriving" i.e., specifying a cosmological model is found at once in the comprehensive unity and logical simplicity of that theory as well as in the general empirical corroboration or successful application it has already received in specially selected areas. Neither ground is, to be sure, compelling, but together they provide the only kind of justification in any case available in any significant alternative. It may be that, ultimately, the theory of relativity will prove not wholly appropriate for handling the growing mass of data accumulated by observational astronomy on its most comprehensive scale. Also, further testing of the theory even in relatively restricted areas might produce difficulties of a serious enough nature to cause a considerable overhauling in the basic structure of ideas in the general principles of the underlying theory. No guarantee against such always-present contingencies can be given. Until such time, however, as these arise, or until in the meantime a more fruitful, i.e., economically simpler, comprehensive, and empirically supported theory makes its appearance and shows its aptitude for making intelligible the facts under consideration, it is perfectly reasonable to continue to use what is at the moment a promising line of attack. Moreover, the kinds of objection against its use to which Milne, for example, has given expression seem not altogether convincing. He deems it inappropriate to apply a theory which has been confined to what he calls "small-scale experiments" (viz., the solar system) as a basis upon which to extrapolate to the universe as a whole—"a grand experiment in being." Now it must be granted that in the construction of the field equations, guidance was obtained from the requirement that such equations "reduce to" Newtonian equations (or its equivalent in Poisson's equation) as a first approximation, and for purposes of initial observational test, predictions were restricted to gravitational situations in a field like that of the solar system. But such circumstances of discovery and initially limited applicability should neither be confused with the logical status of the theory, nor taken as a sufficient ground for judging its possible total adequacy. It would, therefore, be a mistake to assume that schemes of relativistic cosmology rest on conceptual foundations whose usefulness is necessarily *confined* to "small-scale experiments." Cosmological solutions of the field equations are just as significant and appropriate domains of application for the pattern of relations expressed by those equations as any other area of physical fact. The largeness or smallness of the fields to which the equations are applied is a matter of theoretic indifference to the possible scope of the equations and the program of field physics since all are, at least in intention, encompassed with equal interest and relevance.

Let us assume, then, that the adoption of the field equations of general relativity has the kinds of sanction which perforce must attach to any other fruitful mode of attack. The making of further assumptions to narrow the choices and provide for the kind of solutions to the equations that will constitute a cosmological model, must now be examined as to type. These again prove to be dominated by considerations in some cases of a formal character—relating to questions of mathematical simplicity and generality—and in others of an empirical character, resting directly or indirectly on clues and results of an observational sort. Consider the "derivation" or warranting considerations for the kind of line-element (metric) to be introduced into the field equations as a necessary step in establishing a particular model. The predominant share of attention thus far has been devoted to "homogeneous" models, whether these were, as in earlier investigations, "static" ones, or as in recent researches, "non-static" ones. We may characterize in a general way the property of homogeneity as follows: For all observers situated on particles belonging to a collection of fundamental particles representing the constituents of the universe, the results of surveys on the entire collection insofar as it is open to inspection by such observers will be fully equivalent to one another. The equivalence or inter-translatability of such descriptions means that the distribution of particles will be observed to be uniform throughout the collection and holds for surveys made by all actual or hypothetical observers located at any point throughout such a collection. A homogeneous model, when considered from a purely geometric point of view is one of constant curvature, one in which all points and regions are intrinsically indistinguishable from one another. The manner of treatment of the distribution of matter throughout the universe is such that all irregularities of a "local" character in the actual distribution of the galaxies are regarded as "smoothed out," much in the same manner that, in the interest of giving an overall picture of the (spheroid) shape of the earth, one neglects relatively small variations in its surface features such as mountain ranges or even the flattening of the poles. The justification for the selection of homogeneous models for study have been two-fold, mathematical and observational. On the first count, homogeneity permits a relative ease of handling which is absent when assumptions of non-homogeneity enter to complicate the picture. Moreover, astronomical surveys of the distribution of galaxies, revealing it to be on the whole of a uniform character, suggests a similar choice for study. The "sample" of the universe examined, the observable region explored, leads to the expectation that further observations would continue to reveal a similar uniformity of distribution. The sanction thus derived from astronomical observations is clearly of a tentative sort, and the adoption of the "sample principle" to guide theory in no way implies that the decision to take the particular fruits of observations available at a given time is irrevocable. Should future observational discoveries point to a non-isotropic, non-uniform distribution of nebulae, the occasion would be more urgently at hand to investigate the use of non-homogeneous models. Indeed the use of a non-homogeneous model had already, in fact, been proposed by some writers to obviate certain difficulties met with at a particular stage in the use of homogeneous models,[4] although there does not seem to be any general agreement as yet in favor of abandonment of the homogeneous models.

Assumptions made concerning the material composition of the universe constitute another crucial phase in the con-

---

[4] Cf. H. P. Robertson, "On the Present State of Relativistic Cosmology," *Proc. Amer. Philos. Soc.*, 93, 1949, 527-531; G. C. Omer, Jr., "A Nonhomogeneous Cosmological Model," *Astrophys. Journ.*, 109, 1949, 164-176.

struction of cosmological models. Once more, such justification as these possess depend on considerations of formal simplicity, coherence with established physical doctrine, and the support provided by available astronomical or astrophysical observations. The nature of the assumptions made range from those which are shared in common by a whole class of models to those which are uniquely characteristic of a particular model. A characteristic treatment is one which regards the universe as one which is filled with a continuous and perfect fluid to which must be assigned specific values of density and pressure. It is on this basis that schema concerning the mechanical behavior of fluids may be extended to the treatment of the universe as a whole. Special assumptions, however, will have to be made concerning the amount and character of the density and pressure of the matter making up the fluid. Thus, whether both are greater than zero, or only one is, or neither, whether the fluid is to be taken as composed of "incoherent" matter exerting no pressure, or of radiation exclusively, or some mixture of the two, again the way in which the matter and energy are taken as regards their constancy or variability in time, whether, further, in the case of changes of density these are to be taken as reversible or not from a thermodynamic point of view—all these constitute special features that serve to distinguish one model from another. The selection of one or another feature is supported by appeal to such clues as direct observation at a given time provides, or on the basis of the most reasonable interpretation offered by current physical theory.

The change from the construction of static to nonstatic models (while retaining the assumption of homogeneity) was similarly accompanied by considerations of a formal and an empirical character. On a purely mathematical level, justification for the use of a line-element containing a term standing for the "radius of curvature" of space, whose "length" is regarded as being a function of time, consists in the fact that a higher level of generality is reached, even at the cost of further complications. That theoretical cosmology benefits from a purely conceptual examination of mathematically defined possibilities revealed by such generalizations cannot be gainsaid. Again, however, a more immediate and cogent reason of an empirical sort was forthcoming with Hubble's disclosures concerning the red-shift of nebulae. Taken as a sign of recession of the nebulae from one another, it led to the consideration of the "expansion" of space as a way of making the fact of recession intelligible. Further, investigations disclosed the "instability" of the earlier (and original Einstein static model), and so led to the analysis of other models which would exhibit traits of contraction, expansion or both (the last being oscillatory in character).

The "static" models of Einstein and de Sitter were surrendered because of their failure to square with the known facts. The former, which posited a finite yet boundless universe, possessed of a finite density of matter, proved unsatisfactory because of its failure to account for the red-shift phenomenon. The de Sitter universe proved similarly defective because of its failure to make allowance for the recognized finite density of matter in the universe. With the introduction of non-static models, whose mathematical possibility was developed by Friedmann and Robertson, and which found support in the red-shift disclosures of Slipher, Hubble and Humason, a new orientation was provided for the construction of a whole new set of possible models.

Here, for example, an early model of the expanding universe employed by Hubble undertook to fix certain of the possibilities left open by the conceptual matrix for a model of the expanding universe based on general relativity, by resorting to available empirical information. Such information was contained both in the coefficients for certain values as stated in the "velocity-distance" relationship of the nebulae, and in the corrected nebular counts obtained from surveys out to various magnitudes. Calculations based on the use of this model, however, proved to be of a character which was quite unsatisfactory on a variety of grounds and so led to its abandonment. In particular, the model required as a calculated result that the mean density of matter be of a considerably greater magnitude (6×10^{-27} gm/cc) than the estimated value generally accepted at that time (10^{-29}). Again, the model led to a "time-scale" which was much too short ($\angle\ 10^9$ yrs.) to be consistent with other data independently available; the universe turned out to be "younger" than some of its constituents, the Earth, for example, whose age had been estimated by means of independent techniques by Holmes and others. Finally, the model led to the assignment of a degree of curvature to the spherical space used which would make the universe so small that one might claim present telescopes to have already explored a substantial portion of it. This seemed unlikely. These consequences of the model were recognized by Hubble himself to be sufficient grounds for its rejection. In addition, a number of criticisms were made by others, including Heckmann, which cast doubt on the empirical data used in the construction of the model. These criticisms were directed at the acceptability of assumptions made by Hubble in the reduction of the raw data and that were reflected in values occuring in the "empirical" formula used in obtaining the model.

Since then a number of other models of the "expanding universe" have been proposed. Each seeks to provide a more satisfactory representation of the facts and a pattern of inference which may be submitted to observational check. Such models differ among themselves in a number of ways: (1) in the empirical data initially appealed to as a basis for specifying parameters in the schema and as, therefore, a basis of calculation to other empirical consequences; (2) in the geometry employed, that is the type of curvature (whether zero, positive, or negative) assigned to the "space" of the universe as a whole; (3) in the "temporal behavior" of the model, for example, whether the universe may be said to have expanded from an initial singular state, or, on the other hand, may be represented as undergoing endless cycles of oscillation between phases of contraction and expansion, or again, is to be thought of as in a steady-state of expansion (with no singular states in its past or future, or cycles of expansion and contraction); (4) in the values assigned to the so-called cosmological term occurring in one form of the basic field equations. The technical details of these models lies beyond the scope of the present discussion. While most models considered at the present time include in some manner a representation of the homogeneous distribution of matter, its finite density, and the red-shift in the spectra of nebulae, they differ in a variety of other details. While no widely adopted decision in favor of any of the current models is to be noted, it is to be expected that continuing inquiry will at least tend to eliminate the unsuitable ones. Should such inquiry reach a relatively stable halting place with the adoption of a favored model, it will in any case point not to the final disclosure of "the truth," but only to the provisional success had in interpreting the known facts of observation.

Even such a very brief and greatly oversimplified review as the one we have given will have served its purpose if it helps to remind us of the underlying pattern of scientific inquiry and the way in which it is illustrated in the case of relativistic cosmologies. Looked at in its historical development, we recognize successive stages of refinement in the application of ideas under the constant control of facts. The use of the field equations of relativity, itself a broad hypothesis that is supported in some areas of established fact where it proves helpful and enlightening, but for which no antecedent guarantee can be given as to its continuing usefulness in other areas, is taken as a promising instrument of inquiry. In turn, it prompts various subsidiary efforts to make it more directly significant in interpreting the subject matter at hand. Some common assumptions (in addition to the use of relativity ideas) guide the construction of a number of models, e.g., the use of the assumption of homogeneity, not only because of considerations of convenience and mathematical simplicity but more importantly because of the support given by observational astronomy. Where the use of homogeneous models begins to show difficulties, trial is made of non-homogeneous models. Where again, within the group of homogeneous models reference is made to the growing mass of observational data, proposals at variance with such new data are in some cases surrendered. The static models of Einstein and de Sitter are given up in favor of those which describe a universe undergoing expansion while at the same time having a finite density.

The general question which has to be answered in all these cases is this: Once the assumptions are made which embody and make use of such conceptual materials as seem most preferable and such empirical data that are considered at once reliable and revelant, is it the case that calculations can be made resulting in conclusions, which, with the aid of a suitable "dictionary" for translating theoretical terms occurring in such conclusions into their empirically measurable correlates (correlates which are independently warranted), can be brought into sufficiently close agreement with the latter? In such a case the model as a set of rules of inference is considered an acceptable or useful one, otherwise not.

Differences of judgment that are apt to arise here are assignable to a variety of factors. Variations in judgment make themselves felt in terms of the preference given to one conceptual scheme rather than another, sometimes in terms of purely formal considerations like those which are taken to constitute mathematical elegance or simplicity. Again, differences arise in an area particularly such as cosmology, which are not easily resolved, because of the relative paucity and inexactitude of the observational materials. Models are not easily confirmed or disconfirmed. Also differences of judgment arise over whether the "independently warrantable" empirical data used as tests of the model, are in fact adequately interpreted. Since no such data ever actually come raw, that is, without the use of some tacitly employed theoretic scheme belonging to some established branch or other of physics, questions may arise as to whether it would be feasible to re-interpret such empirical materials in a way which would bring it into greater conformity with the proposed cosmological scheme. Thus the appeals to mechanics, quantum theory, spectroscopy, thermodynamics, optics, electro-magnetic theory, are everywhere present in either an explicit or tacit way wherever use is made of ideas like the Doppler effect, the obscuring effects of matter in the galaxy upon the passage of light coming from distant systems, the radiating mechanisms of stars and other luminous bodies, the comparatively negligible pressure of radiation, the corrections which have to be made on photographically obtained magnitudes in allowing for the red-shift and in order to obtain bolometric magnitudes and so on. Such appeals to the "empirical" whether direct or indirect, are constantly subject to modification and correction.

While surely the differences noted cannot be either minimized or overlooked, and in the case of a discipline like cosmology are indeed more pronounced than in some other areas of scientific inquiry, still the situation is not basically different in principle from what it is in other sciences. Progress is to be noted, even if made slowly in terms of a constant jockeying back and forth between theory and facts. Ideas become modified and replaced by those more adequately suited to the explanation and prediction of observed fact. What belongs to "observed fact" however, is itself subject to change not only in terms of additions made affecting the amount and detail of what is known, but also in terms of the interpretation given to the "facts" as observed. Duhem has put the general point with characteristic forcefulness: "When certain consequences of a theory are struck by experimental contradiction, we learn that this theory should be modified but we are not told by the experiment what must be changed. It leaves to the physicist the task of finding out the weak spot that impairs the whole system. No absolute principle directs this inquiry, which different physicists may conduct in very different ways without having the right to accuse one another of illogicality. For instance, one may be obliged to safeguard fundamental hypotheses while he tries to reestablish harmony between the consequences of the theory and the facts by complicating the schematism in which these hypotheses are applied, by invoking various causes of error, and by multiplying corrections. The next physicist, disdainful of these complicated procedures, may decide to change some one of the essential assumptions supporting the entire system. The first physicist does not have the right to condemn in advance the boldness of the second one, nor does the latter have the right to treat the timidity of the first physicist as absurd. The methods they follow are justifiable only by experiment, and if they both succeed in satisfying the requirements of experiment each is logically permitted to declare himself content with the work that he has accomplished. . . . In any event this state of indecision does not last forever. The day arrives when good sense comes out so clearly in favor of one of the two sides that the other side gives up the struggle even though pure logic would not forbid its continuation." [5]

Chapter 6

Rationalism

WE MUST now consider some objections of a philosophical character that have been urged against the general conception of scientific method just developed and specifically against the procedures employed in relativistic cosmology. It will be no part of our intention to defend relativistic cosmology on technical grounds nor to suggest its inherent superiority to other possible theoretical approaches. What rather we must examine is whether the *kind* of approach represented by relativistic cosmology is open to the charges that have been

[5] Duhem, Pierre, *The Aim and Structure of Physical Theory*, 217 f.

levelled against it and that might, therefore, confront other theories which undertake to do the same sort of thing. The focal point of controversy surrounds the question of the manner of validation of theories or models of the universe and the relation these bear to the facts of observational experience. The spearhead of the attack is led by those who uphold a *rationalist philosophy of science.* This philosophy rests upon the generic demand that all systematic expressions of knowledge exhibit a pattern of necessary inference connecting items within the system, in such a way that the premisses from which conclusions are derived be of an intuitively warranted characters, unique and unexceptionable, hence conferring the same traits upon the conclusions based upon them. It is commitment to the ideal implied in such a philosophy of science which is at the bottom of those attempts in recent cosmological inquiry—of which Milne's and Eddington's are the chief illustrations—that seek to supply an axiomatic basis for the hypotheses proposed that would insure their truth even apart from the usual types of observational appeal made in the course of empirical inquiry. It is on the basis of allegiance to this kind of goal that, in particular, Milne proposes to displace the "conventional" method of empirical science operative in schemes of relativistic cosmology. The essential difference between a theory such as Milne's and those of the conventional type is that for the latter it is not thought either necessary or possible to "derive" a theory as a unique or demonstrable necessity from intuitive premisses. Instead, one adopts along with such observational clues as are given, whatever most promising and relevant physical theory is available for the interpretation of such data. No presumption is made that such a theory is in any way certain, complete or unique. Milne, on the other hand, feels that to operate in this hand-to-mouth, tentative, and exploratory fashion is inadequate. The deficiency of the conventional method in his eyes is that it makes the investigation of the properties of the universe as a whole dependent upon the use of laws which have been discovered to hold true in, at best only an "empirical" and "inductive" manner of selected portions of the universe. The laws themselves, however, remain "irrational." Instead he would establish cosmology on a wholly deductive basis as a self-sufficient discipline. He would "study the universe as it actually presents itself to us, without appealing to any small-scale experiments or to knowledge derived from small-scale experiments. That is, in short, to use only such brute facts, such irreducible facts, as are of the intuitive sort or do not rest on the questionable principle of induction and thus to appeal to no empirical "laws of nature" of a quantitative kind." [1]

It will prove instructive to examine the ideas underlying such conception of science as it operates in the field of cosmology—because not only will it lead us to expose the familiar yet radical fault in assuming that there are available rival "deductive" and "inductive" methods in science—but in so doing will reinforce certain well-known yet hard-won lessons of methodology, and thereby help prevent those debasements or backslidings of methodologic understanding that could not but impede the healthy development of cosmology itself. We shall find, among other things, indeed, that the kind of model which Milne himself develops has whatever fruitfulness and significance it does as a scientific hypothesis, only to the extent that it, like schemes of relativistic cosmology, submits to the same evidential requirements and criteria of evaluation as are everywhere employed in empirical science.

1. The Argument from the Uniqueness of the Universe

One form of objection to the type of procedure illustrated in relativistic cosmology is based on the appeal to what is called the uniqueness of the universe. Because the universe is unique, it is claimed, the applicability of the "ordinary" method of science with its allegedly normal "inductive" methods, involving the establishment of laws or generalizations based on an examination of multiple instances of some phenomena or class of objects, must fail. Since there is but one universe to examine and explore, all we can do, it is argued, is either to apply the laws of the behavior of physical systems of lesser scope than the universe itself or else make an independent study of the universe and determine by a study of its properties what its basic features are. It is claimed that the so-called inductive or extrapolative technique is that which is employed in relativistic cosmology, but with questionable results, whereas it would be desirable to proceed in the second direction, in the so-called "deductive" manner. Two quotations may be given to illustrate such claims: "The fact [is]," Milne writes,[2] "we have only one universe to describe and to live in. If we believe the universe to be rational, we must be in a position to give a 'sufficient reason,' as Aristotle or Leibniz would have said, for each feature of the universe that we encounter. That is to say, there must be a unique solution to the question 'Why?' applied to the universe, in respect of each feature. This shows at once that something is wrong with the various solutions of the cosmological problem given by general relativity—expanding, contracting, and oscillating universe models, in flat, spherical, or hyperbolic spaces, with and without the so-called cosmical constant. The reason for the multiplicity of these solutions is that they have none of them begun on a sufficiently primitive level. They have not begun by studying the possible forms of the universe *ab initio,* without assuming all kinds of results taken from laboratory physics, and from conventional and current physical theory. When we want to study so vast a problem and so deep a problem, as the *raison d'être* of the universe, we cannot afford to take over any results from currently accepted physical theory. Starting from first principles (like Descartes) we must pursue a single path towards the understanding of this unique entity, the universe; and it will be a test of the correctness of our path that we should find at no point any *bifurcation of possibility.* Our path should nowhere provide any alternatives. The account of the universe I am about to put forward has this property—that at no point does it give alternatives."

The suitability of ordinary methods of science is similarly questioned by Bondi who writes:[3] "A difficulty peculiar to cosmology is the uniqueness of the object of its study, the universe. In physics we are accustomed to distinguish between the accidental and the essential aspects of a phenomenon by comparing it with other similar phenomena. The laws of free fall were found as the common part of numerous experiments on falling bodies, the laws of planetary motion were inferred from a knowledge of the orbits of many planets. This method of abstraction is so customary in physics that it has become usual to formulate physical laws as differential equations (representing the common elements of many phenomena), the solutions of which are only specified by the

[2] *Modern Cosmology and the Christian Idea of God* (Oxford, 1952) 49.
[3] *Cosmology,* 9, 10.

initial conditions (giving latitude to insert the individual characteristics of each phenomenon). The uniqueness of the actual universe makes it impossible to distinguish, on purely observational grounds, between its general and its peculiar features even if such a distinction were logically tenable. It is this impossibility of direct abstraction from the observations that rules out the usual inductive approach."

To evaluate these claims it will be helpful to distinguish two questions: (1) what *meaning* shall we attach to the assertion that the universe is unique and what type of evidence is there to support it? (2) what *justification,* if any, is there for the claim that because of its allegedly special character, the study of cosmology requires a method essentially different from that employed in other sciences?

(1) What, now, does it mean to say that "the universe is unique" and what kind of evidence is there to support it? Of the senses of the phrase "the universe" that have been distinguished, it will be sufficient to confine our attention to two, namely to what is referred to as "the observable universe" and "the universe" as this is defined on the basis of some cosmological model. Whatever is found to hold as a result of our analysis of "the universe" in these senses holds for any others in scientific cosmology which derive from either of these meanings and hence may easily be extended to them.

One way of expressing what we mean, in general, by saying that something is "unique" is to say that the object or entity to which we are referring is such that there is at least one and at most one of its kind in existence. Contained in this usage there are two important ideas, that of *number*—that there is one and only one object or entity, and that of *kind*—the possession of certain distinctive qualities in the broadest sense of this word. The analysis of what these involve will reveal to what extent the claim that anything is unique is a matter of *definition,* hence arbitrary, and to what extent it is a matter of *experience,* hence not a matter of choice.

In general, what we mean by the "kind" of something is determined in part by a matter of convenience, there being no purely "natural" scheme of classification which establishes the absolute superiority of one conceptual scheme for grouping things above all others. It is always possible, therefore, as a matter of principle, so to choose how we shall characterize the differences among kinds, that what on one scheme of classification will allow for multiple instances, on another

scheme will require that for a given kind only one instance
be permitted. (For traditional theology, God is defined as
unique and for Leibnitz the universe is unique also by definition simply because it is "the best of all possible worlds.")
The sole question to be satisfied in any intellectual scheme
of classification is one of determining what can be accomplished by way of fruitful organization and interpretation of
our experience as a result of putting into practice the given
scheme.

*In scientific cosmology when we speak of the universe on
the level of theory, by definition we always so characterize
the universe that it is to constitute the most inclusive set of
objects.* Whether the language adopted by a given theory
speaks of "the universe" as an all-inclusive "whole," as the
widest "class" (or "collection" or "set") of physical objects,
or as an "individual" or as "a series" of events, is a question
of what is considered to be the most apt and fruitful way in
which to characterize the observational data. In each case,
however, whether "whole," "class," "individual" or "series"
(or any other that may be preferred) as defined by the cosmological theory, the universe will be a *unique* whole, class,
individual or series. The uniqueness of the universe on a theoretical level is *not* established by any appeal to experience.
For the purposes of cosmological theory the universe is *defined* in such a way that no other instances of it *can even be
conceived;* no other whole, class, object or series having a
wider or more inclusive scope is permitted by the theory itself. Thus we cannot ask abstractly on a theoretical level:
"Are there many universes or is there only one?" The question cannot even begin to be discussed until we have established our conceptual frame of reference on the basis of some
particular cosmologic theory. Such a theory, however, will define its own conception of what it is to be "the universe."
From the vantage point of the given theory it makes no sense
to ask about the possibility of "other universes." However,
since there are only the vantage points afforded by successive
cosmologic theories, each with its *own* definition of "the universe," there is *in general no meaning* to the discussion of a
plurality of universes from the point of view of theory.

Another way of expressing the point we are making is to
say that what we mean by "the universe as a whole" is not an
empirical concept. To be sure, as a scientific theory, a given
cosmological model will have to permit specific inferences and

predictions to be made by its means when applied to particular observational materials, and among these inferences and predictions some will have to stand the test of experience, a test by means of which we would undertake to determine the fruitfulness, accuracy, and relative worth of the theory with respect to others. *But among these consequences one cannot find as a matter for empirical corroboration anything which may be identified as the universe-as-a-whole.*

We conclude that to ask a given cosmological theory to allow for other universes can, therefore, either mean nothing at all, as universe is defined in terms of that theory, or else that what we have is an elliptical expression for saying that the given theory is not the only one we may employ. We may abandon a particular theory with its associated concept of the universe as unique, in favor of some *other* theory with *its* own distinctive conception of the universe as a unique whole. Whether or not, therefore, the universe *is* unique, from the point of view of the entire field of theory construction, is better formulated as: whether or not a particular cosmologic *theory* is acceptable. And to this, of course, one can never give any answer which will uphold the claim that one theory is *absolutely* superior to all others actual or possible.

Let us now turn to "the observable universe" and ask whether, or in what sense, one may speak of *its* being unique. By "the observable universe," as we have already pointed out, we understand, in general, the most inclusive set of astronomical objects identified with the aid of available instruments at a given time. At the present time this set is constituted of the swarm of individual galaxies and clusters of galaxies as these are identified with varying degrees of observational refinement out to a distance of several billion light-years. Can we say this set is unique? Taking the set of galaxies at present identified as a single, relatively homogeneous object, class, or system, that there is at least one such object, class, or system is, of course, what is established directly by observation. But what about the claim that there cannot be *more* than one such object or system? Clearly this is not a matter which can be settled by definition, nor is it in any way self-evident. It is wholly open to further observational investigation to determine whether continued probing by more refined and powerful instruments will simply enlarge the range of the collection we now identify and thus continue to leave it as a "single" system, or, on the other hand, will provide grounds for recognizing that the present set of galaxies is but a unit in a still more inclusive "supergalaxy" or some other system of higher rank. The claim to the numerical uniqueness of the observable universe is one which, while supported at the present stage of research, remains nevertheless an open and contingent proposition. And the same conclusion considered on observational grounds alone will always be true of *any* collection, however extensive and whatever its organization as our instruments disclose it.

(2) If the foregoing analysis of what it means to speak of the "uniqueness" of the universe is not altogether wide of the mark, it follows that the attempts made to appeal to such uniqueness as a basis for the claim that a different method from the ordinary one of science must be employed in its study, are groundless. Milne's claims to have established in his own model a theory which follows a strictly deductive path and provides a unique solution will be examined below. His objection to relativistic cosmology on the ground that it leaves open an unwelcome and unnecessary multiplicity of specific models to choose from is without merit. Such openness of possibilities allowed by relativistic cosmology is in fact a virtue rather than a fault. It exhibits the conceptual richness and range provided by the general field equations from which cosmology draws its support. The fact that the framework of ideas encompasses multiple distinctive possibilities but does not by itself indicate which specific one is to be selected shows a fundamental methodological soundness in leaving such choice to be effected by appealing for guidance to observational data. Indeed the claim to be able to dispense with such empirical appeals and to be able to select *a priori* some unique conceptual model is itself the rationalistic dogma to be criticized and guarded against.

The other objection noted above against the use of "conventional" procedures in cosmology that, namely, in dealing with the universe as a unique object, the normal distinction between law and instance no longer holds, hence the appeal to "inductively" established laws in cosmology cannot be upheld, is similarly without force. The reply is clear and may be briefly stated. The objection rests upon a misconception concerning the nature of science. It makes the mistaken assumption that science is fundamentally or even exclusively concerned with establishing laws in the form of empirical generalizations, arrived at by summarizing in inductive fashion the traits found in multiple instances of some domain of natural fact. While such indeed might, with suitable qualifications, be recognized to hold for certain early or "descriptive" stages of science, it surely does not hold of science on its advanced levels. For these it is more correct, as we have seen, to locate the goal of inquiry in the discovery (or creation) and use of comprehensive theories. And theories are not generalizations; they are not arrived at by a process of "induction" from an examination of multiple instances. They are patterns of thought used for the interpretation of facts but not themselves derived from a process of observational inspection. In the domain of cosmology the opportunity to make use of such theories and the resultant value of such use are in no way different from what they are in other areas of investigation.

2. Milne's Kinematic Relativity

As an illustration of an attempt to appeal to rationalistic criteria in setting up a cosmology and the inevitable shortcomings in such an approach, it will be of some interest to examine briefly Milne's own constructive efforts. These efforts culminate in a cosmological model that is based on a doctrine called "kinematic relativity." [4] It was devised in an attempt to give an explanation for the phenomenon of the "expanding universe" that would avoid appeals to the notion of a curved expanding space and the use of the dynamical and gravitational concepts involved in the field equations of the general theory of relativity. For Milne the *suggestion* for the model he develops was found in the phenomenon of the nebular red-shift and its natural interpretation as a case of recessional velocity. However, the *justification* for the "derivation" of the model, in his eyes, is in no way initially dependent upon such empirical, observational fact. Rather it rests on what he takes to be the inevitable, necessary premises which express intuitively what is involved in the very making of measurements and the way in which intercomparison of measurements by a number of observers would have to take place if any coherent scientific results were at all possible. This leads him to the analysis of what is taken to be the most fundamental of all measurements, that of time, and the development of a principle of relativity couched essentially in terms of the re-

[4] Cf. Milne, E. A., *Relativity, Gravitation and World Structure*, and *Kinematic Relativity*.

lation of equivalence in such time-measures. The absolutely primitive fact upon which both cosmology and physics are made to rest is the experience of the temporal "before and after" relation in the consciousness of a single observer. Such an observer is assumed capable, in the light of this unequivocal type of experience, of ordering all events that occur "at" him in the serial order of "before and after." Whenever, Milne proposes, the observer establishes for any sequence of events at himself a correlation, however arbitrary, between the order of those events and the linear, monotonically increasing sequence of real (or even complex) numbers, a "clock" has thereby been set up. A clock for Milne is thus primarily a scheme for the enumeration of the order of events rather than an instrument for the measurement of the duration of intervals. (It is here that his procedure diverges in an important respect from that which is commonly employed in the definition of clocks in conventional physics.[5] For, strictly speaking, all that he makes use of are the properties of transitivity and asymmetry in the ordering relation, not what is essential for measurement in the full sense, namely a conventionally chosen basis for defining the *additive* properties of time-intervals.) Out of the "measures" (in Milne's sense) obtained by a single observer with the use of such an arbitrarily graduated clock, the observer can proceed to obtain other "measures"—for example of the distance and epoch of events that occur in his environment. The situation may be broadened to include a number of different observers, separated from one another, each possessed of his own clock, arbitrarily graduated, and each accordingly obtaining with the use of his own clock, measures expressing the epochs and distances of events. Under what conditions would there be agreement in the description of commonly observed events among these several observers? Here Milne characterizes as the essence of a principle of relativity the requirement that, provided these observers employ methods of measurement that are basically alike, they will obtain results in the form of descriptions of the field of phenomena that are identical in form. The formulations made by one observer will be capable of translation into those other observers such that the common content of their accounts will be apparent. For such a community of description to be possible, the essential prerequisite is the possession of congruent clocks by the several observers. While each observer initially may have some arbitrarily graduated clock, through inter-communication (by means of light signals) and comparison of measures obtained by the several observers, all clocks may be regraduated so as to yield a uniform basis of description. When this is accomplished, the community of observers constitutes an *equivalence*. When such an equivalence is further qualified in such a way that the density-distribution of particles is homogeneous, namely, such that the account given of the density-distribution of the collection as a whole in its temporal unfolding is one which is given in the same manner by all members of the collection, there being no preferentially situated particle-observer, the collection then becomes a *substratum*. It is this substratum which Milne proposes to identify with the universe, the particle-observers being the galaxies themselves. According to one mode of clock-graduation, that which for Milne is the more fundamental because of its alleged "pre-eminent, simplicity," [6] the collection of particles forms a uniform-motion equivalence which came into existence as a point singularly at $t = 0$ (creation) and from then on has been in a state of expansion. Associated with this time-scale there is the use by all particle-observers of a geometry having Euclidean properties by which to describe the collection as a whole. It is possible to reinterpret the features of the substratum by employing another scale of time ("the τ-scale") rather than the basic "t-scale" which has the effect of transforming the description of the universe from a swarm of receding particles to one in which they are stationary with respect to one another. There is then no finite epoch of creation—the past becoming infinite, and the space within which the galaxies are situated is then taken as hyperbolic and infinite.

Without presuming in any way to judge the technical adequacy of Milne's model, of which the foregoing only provides the barest hint of its detailed structure, it may nevertheless be pointed out from a purely methodological point of view that, like all rationalist ventures, Milne's effort runs aground with respect to the fundamental assumption that its self-evident beginnings are really self-evident. To begin with, it has been

[6] *Modern Cosmology, and the Christian Idea of God*, 56.

shown by Robertson [7] that, even if with Milne one postulates an "operational" basis for observers consisting in the use of clocks, theodolites, and light-signals, and posits a homogeneous distribution for such particle-observers, for such restrictive conditions a general metric of precisely the degree of generality employed in general relativity can then be derived. Such a metric, however, cannot be further uniquely determined on the basis adopted among the choices it leaves open. In particular the decision as to whether the geometry of the three-dimensional space component of the four-dimensional space-time metric has a (constant) negative, positive, or zero curvature is one that must be left to other considerations to help determine. Milne's assertion, therefore, that the substratum in its most fundamental manner of conception is one in which "private" Euclidean spaces are to be employed to describe the expansion of the collection from an initial singularity occurring at $t = 0$ is arbitrary rather than logically forced on the operational basis he himself adopts.

Again, how does Milne justify the introduction of the assumption of homogeneity in his model, the substratum? He would have us distinguish the situation with respect to the substratum from that which holds in relativistic cosmologies, for, he claims "the postulated homogeneity of distribution of nebular nuclei—the postulate that an observer at each such nucleus has the same view of the distribution of matter-in-motion in the universe—is not of the nature of an experimental or observational assumption at all. It is of the nature of a *definition*, a definition of the type of system it is proposed to consider." [8] Such an account, however, is perplexing; for if we examine what Milne might mean here by a definition, it turns out to be equivocal. If the study of cosmology were a wholly mathematical one, then, of course, the investigator has the perfect liberty to select any possible scheme of ideas for study, and the selection of postulates, axioms, and definitions is a matter governed entirely by considerations of internal consistency and formal mathematical interest. In such a case, however, one is at liberty to select *any* assumptions that are compatible with one another, hence *no special preeminence attaches to that of homogeneity*. Non-homogeneous models are as perfectly legitimate constructions of mathematical intelligence as homogeneous ones. If, however, one considers the definition of homogeneity not as a mathematical postulate but as part of a *theory* having empirical applicability, then along with such a theory it has whatever sanction the available but incomplete and tentative observational data provide. It is always possible that changes in such data will take place to cause abandonment of homogeneous models altogether. To appeal for guidance to the disclosures of observational astronomy as Milne does on those occasions when he not only indicates the genesis of his own inquiries, but also appeals to it as a field of possible application for the mathematically contrived model which he creates, is to squarely subordinate the abstract but hypothetical necessities of the mathematics to the contingencies of observed fact. Either, then, we have for homogeneity treated as a definition, a mathematical postulate or part of an empirically functioning theory. In either case it is not something for which one can claim self-evidence. Possibly it was the uneasy sense of this dilemma which made Milne on at least one occasion appeal to a rather desperate expedient for rescuing the "inevitability" of his premiss. He writes: "Whether the universe may be *expected* to be representable by a system satisfying the cosmological principle [his name for homogeneity] is a metaphysical question . . . My own private opinion is that the universe must satisfy the cosmological principle, because it would be impossible for an act of creation to be possible which would result in anything else." [9] What private sources of knowledge Milne or anybody else possesses for knowing what the universe may be *expected to be* (whatever that might mean) or why one must *assume* to begin with that the universe was *created*, all these remain mysteries to which surely no readily available satisfactory *scientific* answers are forthcoming.

When we turn, finally, to the other main pillar of support for the justification of Milne's model, that which consists in exhibiting the fundamental character of time-measures as it determines the whole pattern of more complex ideas, there are two divergent ways of looking at the significance of what he has accomplished. The goal of providing in a theory systematic organization for the multiple facts of a particular domain is everywhere guided by the desire to achieve such

[9] *Relativity, Gravitation and World-Structure,* 69.

organization with a minimum of basic ideas, that is, with logical simplicity and economy. We admire the classic simplicity of Newtonian mechanics and the way in which it is established with the aid of the basic ideas of time, length, and mass. A similar criterion of evaluation may guide us in finding in Milne's ingenious and extensive studies an attempt to show the range and ramifying consequences of the concept of time, as he interprets this, for physical understanding. Yet however suggestive this effort may be, one cannot claim for it any finality or exclusive priority over other selections. Efforts at simplification must always be regarded in the context of a total theory and the capacity such a theory has for applying successfully to empirical materials. This cannot, however, be decided in advance in terms of the particular conceptual or axiomatic base chosen. Moreover, in the very use of *any* theory, Milne's included, there are innumerable lines of connection with a funded body of empirical knowledge and the conceptual materials deriving in part from the accepted use of *other* theories. For example, in practice, should Milne's model of time-measuring observers be applied now in the interpretation of actual observational materials in the domain of cosmology, many references to the accepted body of physical knowledge would have to be made. Consider, for example, the use of clocks by members of what Milne calls an equivalence. Milne recognizes that instead of requiring reference to the "consciousness" of observers attached to galaxies, we fall back on the "clocks" provided by natural phenomena themselves. One such clock is taken to be the atom. It is the radiation of such atoms, received as light rays, which, in terms of noted frequencies, is used as a measure of time-keeping. The light received from distant galaxies contains important information for incorporation in the cosmological scheme. However, reference to "light" or to "galaxies" or to "atoms as radiating mechanisms" depend for their meaning upon an implied acceptance of an extensive body of empirical data and theoretical interpretations (other than purely cosmological ones); and clearly there can be no question raised which doubts the incompleteness and corrigibility of the areas of knowledge touched on by such references. This has the consequence that an applied or empirically functioning cosmological theory is incapable of being justified in any intuitive or demonstrative way.

Chapter 7

World Geometry

IN THE construction of a cosmology, spatial terms play an important and prominent role. Traditional accounts centered on efforts at deciding whether the space of the universe is infinite or finite. In the present day discussions one finds references in addition to such ideas as the "curvature" of space and its "expansion." With respect to these matters there are three questions that may be raised: One has to do with the meanings to be assigned to such terms as "curvature," "infinite," "finite," "expansion" and the like. For the satisfaction of this interest, a knowledge of the language of geometry is required and in addition of such specifications of usage as is provided by the physical theories in which they figure. A second question relates to the evidential grounds for effecting a choice among the various claims made with respect to the spatial properties of the universe as a whole. For a satisfaction of this interest, a detailed examination and assessment of the evidence upon some commonly agreed to standard of what constitutes satisfactory evidence would be necessary. A third type of question relates to the logic of such statements. It seeks to determine what *sort* of thing we propose or claim to have established when we say that the universe has one or another of the above specifically spatial characteristics. About *what* are we talking? What kind of information or insight, what type of cognitive value is contained in statements of the sort mentioned? What kind of evidence or criteria *should* we appeal to in making a choice?

This last type of problem, one with which the following discussion is concerned, may be illustrated by making reference to two topics which have been widely debated. One has to do with the "reality" of physical space, the second with the question as to whether or how its specific metrical structure is to be determined. These two questions are sometimes lumped together in the query "Is space really curved (or expanding)?" Controversy and differences of opinion abound, not because of any failure at common understanding of the technical geometric meanings involved, nor even because of any primary difficulty in judging available evidence, but because of radical oppositions or obscurities connected with logical issues. These differences and obscurities center on such questions as whether space may be spoken of as some type of objective entity or structure and how this is to be understood, or again, whether an appeal to experience is sufficient to isolate from the several metric possibilities the one which is said to "hold true in fact" and what it means to say this.

1. Concepts of Physical Space

The concept "space" has a great variety of meanings, as is made evident from both a history of the subject as well as from a survey of the usages it enjoys at the present time. It is a concept which is variously understood and employed in the domains of ordinary experience, psychology, the practical arts, pure geometry, and physics. Our interest in the present discussion is in "space" as it is understood in physics generally and in cosmology in particular.

As Einstein has pointed out, we may recognize and trace the parallel development of two fundamental views of space which have played important roles in the history of physical thought.[1] The concept of space which Einstein himself favors, as we shall see, cannot be identified in any simple way with either of these traditional views. Rather, it incorporates important elements of both and offers us, therefore, a third distinctive outlook. The two earlier views have their source in distinctions already recognized on the level of primary, everyday experience. They are given a more elaborate and sophisticated expression in the philosophic and early scientific views of Greek thought. In the modern period, with the birth of science, both make their reappearance. However, the incorporation of one of these modes of approach in the overwhelmingly successful theory of Newtonian mechanics has tended to make it dominate discussions of the subject, at least in physics, until the advent of relativity theory.

On the level of primitive thinking as well as in the majority of classic philosophic systems, a central role is played by the idea of "objects," "things," or "substances," as providing the basis of organization and interpretation for the manifold succession of sensory experiences. Where with further differentiations the notion of *material* objects makes its appearance,

[1] Foreword by Albert Einstein in Jammer, Max, *Concepts of Space* (Harvard, 1954) xii ff.

an important concomitant set of devices for the ordering of our experience of such objects is found in the use of spatial and temporal terms of description.

Broadly speaking, the origin of the concept of space lies in the twin needs of providing some basis for specifying the position *where* material objects are located and a *medium through which* they may be said to move. Each demand has given rise to its own typically oriented conception of space, which, as it becomes elaborated, undertakes to encompass the other within its own framework of ideas.

According to one typical demand for spatial representation, we seek some way of indicating where an object is to be found, its position with respect to other objects. This gives rise to the concept of *place*. It also underlies associated efforts at specifying the size and shape of objects and their mutual distances from one another. Eventually, it brings with it the development of elaborate schemes for identifying locations such as are involved in the use of the many types of co-ordinate systems of reference. This is what we may designate as the *"positional"* conception of space. It encourages a *relational* interpretation of space, rather than a substantial or quasi-substantial one. Space becomes a network of relations among bodies or material objects. Without the latter there would be no space. The Aristotelian conception of space as "place" belongs essentially to this positional-relational approach.

Meanwhile, another conception of space is developed out of the interest in representing those features of the arrangement of material objects which have to do with the phenomenon of *displacement*. Where an object was previously located, another object may come to be situated. The place remains. Not only can objects replace one another as in a container, it appears natural to endow the "opening" or "interval" left when "nothing" is there with some positive reality of its own. Here is the origin of Plato's metaphor of space as the "womb" in which the world is born. One now refers to the space in which things may be put. It is also easy to extend the idea and think of this same space as that in which they move. This *"container"* conception of space submits itself to various modes of refinement and complication. Whereas in its primary context of illustration every container is bounded by sides that are material, one line of development, through a process of conceptual idealization and abstraction, undertakes to push back the sides to ever-increasing volumes at the same time that—through an analogous limiting process—the sides themselves are reduced in thickness, so that, "in the limit" both the ever-receding sides and their thickness disappear. Finally, another idealization may take the form of conceiving of all matter, however diaphanous, as being removed from such a container. Thus emerges the notion of space as a limitless void —as we meet it, for example, in the thought of the ancient atomists. Material objects and space now become *two* primary realities. For such a view in general, space, whether or not it is given some type of ontological priority as the "receptacle" in which bodies are placed and in which they move freely, at least enjoys a status of independence with respect to bodies. Space, no longer the relational pattern of bodies, becomes the medium in which bodies are placed and displaced. It makes sense on this view, as contrasted with the other, to think of space apart from matter. It was out of such germinal ideas of space as a container medium, the setting for the motion of bodies, that the concept of absolute space was developed and made to serve a special role in Newton's system of mechanics. It was due to this fact, and the prestige which this system enjoyed, that the "container" conception of space dominated physical theory from the seventeenth century to the advent of relativity. Despite the numerous objections raised by philosophers like Leibnitz and Berkeley, the absolute conception of space overshadowed a relational view only because it was part of a system of physics which gave extensive and precise comprehension to phenomena. Whatever uneasiness with the idea of absolute space was felt, even by Newton himself and by other physicists, could be submerged in light of the evident success which the system of mechanics had as a whole. The criticisms of the philosophers —and they were in their own way sound enough—could not enjoy the influence they deserved among practising physicists without a comparable embedment in a positive physical theory; and this was lacking.

What was it that Newton added to the "container" conception of space that made it an indispensable part of his physics? Newton needed absolute space as a frame of reference by which to account for the presence of inertial effects that make their appearance in the accelerated motion of bodies. As the famous bucket experiment was intended to show, it was insufficient to appeal to the relative motion of bodies with respect to one another in order to account for the presence of inertial or centrifugal forces. It was necessary to assert the inertial effect of space upon such bodies in order to account for the observed effects. Space was thereby given a uniquely privileged status. In being an active causal agent producing inertial effects it was not reciprocally involved in any dynamic interaction; *it* could suffer no effects. As Einstein has pointed out, it was far from being a mere addendum on Newton's part to include the notion of absolute space in his mechanics. It helped, rather, from a logical or systematic point of view to make his physics consistent and complete. To object to it on the ground that it is empirically unverifiable, that it is, from an operational point of view meaningless, is to miss the whole point. In fact neither the relational point of view nor the relativistic field conception, nor indeed *any* scheme of spatial interpretation as it figures in a physical theory is, as such, empirically or operationally identifiable in any simple or direct way.

Mach's well-known criticisms of Newton's views were intended to bring about a greater unification in that system of ideas by requiring that inertial effects be defined not with respect to space but with respect to bodies. Let the fixed stars, Mach suggests, be used, for example, as the frame of reference for the rotating bucket instead of space. What holds with respect to accelerated motion can then be brought into conceptual continuity with translational motion, as always involving relative motion of *bodies*. Mach's views, suggestive as they were, could not bear fruit until taken up and integrated within a scheme of physics which could in practice prove its superiority to the Newtonian system. This was not accomplished until the advent of the general theory of relativity; and even here, Mach's ideas underwent important modifications before they could be made to serve the purposes of a field physics.

The "container" conception of space had still another independent, major source of support in nineteenth century physics, one that came to it from the development of electromagnetic theory and its incorporation of optics. Through the work of Maxwell, Hertz, and Lorentz the idea of absolute space was retained in the form of the ether. A quasi-substantial medium composed of "imponderable" matter was conceived as filling the absolute space of Newton to serve as the means of transmission for radiation and as the carrier of the field. In such views, although the field now makes its appearance as a major conceptual device in place of the appeal to particles in motion, it is, nevertheless, under the pressure of such older views, still subordinated to the notion of matter. The field must be sustained either by ponderable or imponderable matter. What made the ether as thus introduced into physics an unsuccessful idea was not its use when confined to the domain of electromagnetic phenomena as such. Again, on logical grounds, it was unexceptionable. It was rather the *combination* of ideas derived from the older mechanics, in particular the classical theorem for the addition of velocities, or, more generally, the Galilean transformation equations, and the postulate of the constant velocity of light as referred to a fixed, space-pervading ether, which, together, were incompatible, in a pragmatic sense, since they led to predictions that were disconfirmed by the Michelson-Morley experiment.

It would be a mistake to think that, at least as far as Einstein is concerned, the conception of space of the theory of relativity involves the abandonment of the "absolute-container" view in favor of a return to the "positional-relational" view. The conception of space rather, that the program of relativity physics envisages is one that offers a third, distinctive, possibility although some elements of continuity with both traditional approaches are retained. The program of relativity physics, one that is still essentially incomplete, is grounded in the expectation of the conceptual sufficiency of the idea of the *field* as a unifying basis, for the interpretation of all physical phenomena. The successes of the special and general theories of relativity are to be looked upon as but steps along the way to the realization of this goal. The role of spatial concepts in particular, from this point of view, must be seen as one side of the working out of an all-encompassing field-physics. Such a physics, however, involves one crucial point of difference from traditional approaches. For *both* classical conceptions had one assumption in common —they started from the unquestioned acceptance of the independent existence of material objects and sought to provide a spatial context for their existence. One might say that on the container view material objects had external relations *to* space and on the positional view that space itself was one species *of* such external relations. Space either as the medium-container for material objects, or as the network of positional, metrical relations among bodies was, in both cases, accorded a reality distinct from the bodies themselves. It is this distinction between container or pattern, on the one side, and content, on the other, which is rejected in the outlook of a field physics. It would substitute the single notion of a field whose variations constitute matter or energy and where the latter, therefore, take the place of independent particles or singularities in the field. Space under such conditions would be a name for the *metric structure of such a field*.

The special theory of relativity as the first step in the realization of this objective, while revolutionary in certain respects, retained some features of the older schemes, features that were to be eliminated in turn by general relativity. The special theory of relativity in giving a successful interpretation of the Michelson-Morley experiment, demanded the rejection of the notion of an ether as a universal imponderable medium filling all of absolute space. To account for the observed facts, however, it was necessary to adopt a new set of transformation equations (those of Lorentz) to replace the classical Galilean ones. With the Lorentz equations it becomes possible to extend the principle of relativity already contained in classical mechanics—this time to *all* laws of physics. However, due to the incorporation of the constancy of the velocity of light as one of the invariances holding for inertial systems (co-ordinate frames in uniform relative motion to one another) it no longer becomes possible to adhere to the classic conception of absolute simultaneity. The classical separation between a three-dimensional spatial framework for physical objects and events, and a one-dimensional temporal framework, is abandoned. What takes its place, as Minkowski showed, is a $3 + 1$ dimensional continuum ("space-time") or four-dimensional "space" for short, which is variously "broken-up" and used in the descriptions of events by observers attached to different inertial systems. However, this four-dimensional continuum, the Minkowski "space" of special relativity, functions conceptually, at least, as did the various forms of absolute three-dimensional space, including the ether. It may be thought of as providing a "medium" or "carrier" for matter and field at the same time that it enjoys an independence of such content or filling. This is made manifest by the fact that even in the absence of matter and field, this continuum has a structure of its own. For, according to special relativity theory, the metric of space-time has a

Euclidean structure which is simply an extension of that which is found in ordinary three-dimensional space. Such a rigid and fixed Euclidean structure is one which, in special relativity as in classical physics, may be thought of as independent of any material content.

With general relativity, however, even these points of similarity between the "space" of special relativity and the classic systems are abandoned in favor of a more radical view. For one thing, it makes no sense, according to this theory, to speak of space apart from the field which is now used to represent physical phenomena. In particular, gravitation is now interpreted as one aspect of the metric structure of the field. With the abandonment of the traditional distinction between inert and gravitational mass, gravitational phenomena become assimilated to inertial ones and both are exhibited as capable of being represented by departures in the metric of the field from the pure Euclidean case. The latter retains its validity only as a limiting condition in *infinitesimal* regions where the curvature of the field may be regarded as reduced to zero. The adoption, as in classic physics and special relativity, of a Euclidean metric becomes no longer valid for any *extensive regions* of space. Instead of a fixed and rigid framework, the opportunity now arises for recognizing variations in the curvature of space or, what is the same thing, the use of non-Euclidean criteria of measurement and calculation in different parts of the field as a whole, in accordance with the variations in the density of matter or energy. Indeed the latter quantities, although represented in the field equations of general relativity by separate tensors, are, in terms of Einstein's programmatic goal, but schema to be replaced eventually by terms which will represent for the "matter" or "energy," states of the field itself. "The field of a material particle may the less be viewed as a *pure gravitational field* the closer one comes to the position of the particle. If one had the field-equation of the total field, one would be compelled to demand that the particles themselves would *everywhere* be describable as singularity-free solutions of the completed field-equations. Only then would the general theory of relativity be a *complete* theory." [2]

Apart from the field, then, there is nothing, and contrary

[2] "Autobiographical Notes" in *Albert Einstein Philosopher-Scientist*, Library of Living Philosophers (ed. Schilpp, P. A., Evanston, Illinois, 1949) 81; cf. *ibid.*, 75.

even to special relativity, not even "empty space." In this sense the field in Einstein's view replaces as a unitary conception both matter (whether ponderable or imponderable) *and* space. It emerges as the true absolute, being indeed not only independent of any other physical existence, but in fact all-embracing. On the other hand, "space" must be seen as embedded in, and as an aspect of, the total structure of the field. In this sense the field conception continues the "relational" tradition and, as Einstein has noted, may be taken as fulfilling the Cartesian goal of identifying extension and body. Also, in its own way, it fulfills and offers a constructive embodiment for Mach's criticism of the Newtonian tradition.

2. The "Reality" of Space; Einstein's views

Does space really exist? This vexing and recurrent question runs parallel to the development of the conceptions of physical space and continues to provoke controversy. An examination of Einstein's views will take us to the heart of the issues here and will serve to introduce the topic.

In opposition to all varieties of pure empiricist philosophies, Einstein has insisted that the conceptual component in all schemes of knowledge is not derivable by any process of logical inference or inductive summation from what we directly sense or observe. Already on the level of ordinary common sense knowledge, and becoming increasingly more noticeable in the domain of science, that which is used for the ordering and the rendering comprehensible of sensible experience is provided by the mind. It is to the free inventions, the constructive and creative capacities of the mind that we must look, if we would understand in particular the methods of science. The very concepts of "material objects," "space" and "time," in whatever forms these assume in specific theories, or at various stages of the development of knowledge, are major instances of this constructive activity. Particularly in the familiar widening gap between the highly intricate mathematical symbolism of theoretical physics and the directly experienced sensory manifold, one finds confirmation of the separate origin of these two components of knowledge, the conceptual and the empirical. Yet Einstein insists, of course, that the effectiveness of the conceptual scheme must ultimately be determined not by any purely rational criterion but by its capacity for "applying" to what is given to us in sensation. From among the proliferating richness of the mind's inventive genius, science selects those schemes only which meet the test of empirical corroboration.

What Einstein's views on the "reality" of space are, must largely be construed in terms of his position with respect to theoretical constructs generally. It is here, however, that one is confronted by a certain difficulty in determining what his views actually are at a crucial point in their development or articulation. The difficulty may be illustrated in connection with the different ways in which we may interpret what the "application" of ideas to experience means. The problem may be put for the moment in historical terms, that is, in terms of Einstein's broad sympathies with both Kant and Spinoza. Einstein's Kantianism, it is true, extends only as far as the recognition accorded to the mind's constructive role in its use of ordering schema for the sensory manifold. It stops short of any insistence on the *synthetic a priori* structure assignable to some exclusive, fixed, and unique pattern embedded in the very constitution of the human mind generically, as in Kant's identification of this structure with Euclidean geometry or Newtonian mechanics. Einstein, like Poincaré in this respect, stresses the diversity, the flexible character and inexhaustible richness of conceptual schemes that are in principle available for ordering experience. Were he strictly to adhere to this type of philosophy, however, Einstein would see in not only all other theoretical efforts in science but in his own as well, at best, instruments for the ordering of the observational data which can in no way claim either finality or the guaranteed promise of continuing success. The "application" of theories to the description of empirical materials would be judged in terms of their *relative* effectiveness for articulating the relations of events as judged by their economy, comprehensiveness, and predictive power. Einstein himself has in many utterances stressed exactly these things. In terms of this approach, however, the view of space as part of the embracing notion of a Field remains *one* possible way of satisfying the physical problem of specifying and interpreting the placement and displacement of objects or centers of energy. It may, indeed, prove superior to the traditional "container" conception or to the "relational" conception, but like them *or any other* conception of space it will have to be judged as part of a total physical theory. If we should choose to speak of the "reality" of space, this can only mean the relative effectiveness or superiority possessed

102 / Space, Time and Creation

by a given theory with its own view of space as a way for representing and organizing our data. The test of "application" would not be one of correspondence with some supposedly independent, existent structure *in rem,* but a pragmatic one.

Now Einstein combines with his "Kantianism" a sympathy for Spinoza. To this side of his thought belongs a realistic and rationalistic faith in the determinable structure of reality. Reality is conceived not only as independent of thought but as possessed of a specific essential structure accessible to rational comprehension. Einstein's rationalistic realism is expressed in the conviction that among the many possible creative conceptual schemes used in the interpretation of experience, there is *an* ontologically sanctioned basis for choice. Nature may be expected to disclose her "secrets" to that scheme which is unique in its superiority to all others. He writes: "The essential thing is the aim to represent the multitude of concepts and theorems, close to experience, as theorems, logically deduced and belonging to a basis, as narrow as possible, of fundamental concepts and fundamental relations which themselves can be chosen freely (axioms). The liberty of choice, however, is of a special kind; it is not in any way similar to the liberty of a writer of fiction. Rather, it is similar to that of a man engaged in solving a well designed word puzzle. He may, it is true, propose any word as the solution; but there is only *one* word which really solves the puzzle in all its forms. It is an outcome of faith that nature—as she is perceptible to our five senses—takes the character of such a well formulated puzzle. The successes reaped up to now by science do, it is true, give a certain encouragement for this faith." [3] Again, he says: "Our experience hitherto justifies us in believing that nature is the realisation of the simplest conceivable mathematical ideas. I am convinced that we can discover by means of purely mathematical constructions the concepts and the laws connecting them with each other, which furnish the key to the understanding of natural phenomena . . . Experience remains, of course, the sole criterion of the physical utility of a mathematical construction. But the creative principle resides in mathematics. In a certain sense, therefore, I hold it true that pure thought can

[3] "Physics and Reality," *Out of My Later Years* (London, 1950), 64.

grasp reality, as the ancients dreamed." [4] It is clear that according to such a philosophy, the "application" of theories is no longer simply a matter of *relative* effectiveness. The choice in principle among alternative possibilities is guided by the faith that corresponding to the many proposed solutions, *one* will be correct. It is this faith which supports Einstein's own prolonged efforts in behalf of a comprehensive field physics. From this point of view, one must see in his espousal of this program and its associated conception of space not simply one encouraged by great successes already had and a relentless drive for logical unity and simplicity. It expresses rather the conviction that nature possesses a specific pattern and the field physics program promises the most likely path for expressing it. Einstein speaks of "this total field being the only means of description of the real world. The space aspect of real things is then completely represented by a field, which depends on four coördinate-parameters; it is a quality of this field. If we think of the field as being removed, there is no "space" which remains, since space does not have an independent existence." [5] When, therefore, one considers on this view what the reality of space comes to, it must be identified with an ontological counterpart to that uniquely superior account from the whole range of actual and possible thought, a counterpart which antecedently and independently already exists in fact.

It is here that one finds the breeding place for the fallacy of reification. The location of the fallacy is to be found in the assumption that Nature is like a *puzzle* (Einstein's own expression) instead of, as it is for science, a series of *problems*. To assume that it is a puzzle, suggests that there is only one correct way in which the pieces fit together, and that it is the task of science to discover that way. But this whole way of talking, it must be insisted, rests on a projection into Nature of what holds only for certain limited areas of human experience. Puzzles, after all, are constructed *by* men *for* men, and we know in advance by virtue of the fact that they are so constructed, that there is *a* solution. We legitimately look for it. But we have no reason to speak of Nature as a puzzle. *We* experience *problems* when we seek to order our experiences and we solve our problems when we get satisfactory

[4] "The Method of Theoretical Physics," *The World as I See It* (London, 1935), 136; cf. "Autobiographical Notes" *loc. cit.*, 81, 667.
[5] *Meaning of Relativity* (Princeton, 1953), Append. II, 4th ed., 163.

ways of interpreting the data that poses them. But this is a relative matter. For comparative successes need not be graded on a single scale on the assumption that there is a supremely satisfactory solution.[6]

3. The Metrical Structure of Space

Our analysis thus far has been concerned with space in a general sense as it figures in physical thought and the question of its "reality." Discussions in physical geometry since the middle of the nineteenth century have also revolved about another, more specific topic, continuous with the first, namely, the methods to be used for determining the particular *metrical structure* or type of curvature of space. This discussion falls, historically, into two stages. At first, the occasion for such a discussion was stimulated by the establishment of the subject of non-Euclidean geometries as accredited parts of mathematics. The axiomatic or synthetic approach to the systems constituting the plurality of geometries established not simply the formal validity belonging to each of these geometries, but the fact that as formal systems, they are mutually consistent and translatable into one another, though not identical. Through the work of Gauss and Riemann the differential or analytic approach to the varieties of metric geometries established the way in which such geometries may be formulated in terms of the idea of curvature. This idea was elaborated by recognizing the means for determining the "intrinsic" curvature of space, the distinction between spaces of constant and variable curvature, and the proof that there are only three spaces of constant curvature, namely, zero, positive, and negative, corresponding to the systems of Euclid, Riemann, and Lobachevsky. The importance of the notions

[6] In a private conversation with Einstein, the author enjoyed the privilege of receiving his comments on the material of the foregoing section. In substance, Einstein's view was that the antithesis claimed by the author was not a genuine one. What is alluded to in the text as his "Spinozism" he regarded as a statement of his "religion" as a scientist. To carry on the work on the level and with the persistence and scope of his own interests, he argued, a faith in the unique intelligibility of the cosmos was required to provide the necessary incentive. Einstein felt that the alternative "neo-Kantian" view—that science is an affair of multiple, pragmatically evaluated symbolic systems—would not suffice, at least psychologically, for this purpose. It would seem, however, to the present writer, without denying the *psychological* differences involved (themselves perhaps only reflections of deeper socio-cultural and historical factors) that these are not adequate to remove the *philosophic* incompatibility between the views discussed.

of congruence and free mobility for identifying the spaces of constant curvature also came to be recognized.

With the realization that, considered as deductive systems, the geometries of Lobachevsky and Riemann are as permissible, indeed as consistent as Euclidean geometry, the question arose as to which of these abstract, possible schemes of metric relations was "actually embodied in the world of fact." Another way of putting the question was: "of the various systems elaborated in pure geometry which one holds true for a given domain of physical phenomena?" In the first stage of the discussion, no guidance in the form of an explicitly developed physical theory for undertaking such an inquiry was present. More accurately, such physical assumptions as were *tacitly* appealed to, were grounded in the commonly accepted classical mechanics and optics. The crucial relevance of such assumptions, however, as affecting the ultimate outcome of such an inquiry, was particularly noted by Poincaré, whose writings on this topic mark a turning point in the development of the subject. It was especially the way in which the general theory of relativity made the determination of the curvature of space an integral part of its own outlook, which together with Poincaré's methodological analysis, may be taken as ushering in the second and current stage of the discussions. With this theory, the possibilities of changes in the degree and type of curvature over extensive regions of the field and in the course of time, become physically significant. In cosmology, in particular, the development of models with one or another of the spaces of constant curvature becomes a standard feature of the various efforts made to arrive at a satisfactory theory.

In both stages of the subject, one finds a widely held thesis that an appeal to observation and experiment would suffice to determine which one of the mathematically possible geometries holds true of physical space. This empiricist viewpoint counted among its adherents, in the earlier period, such figures as Gauss, Schweikart, Riemann, Helmholtz, K. Schwarzschild and others. Defense of the same broad thesis, now within the framework of relativity theory has, in more recent years, been made by writers like Reichenbach and Robertson. Such empiricist claims, particularly in their more recent subtler forms, though incorporating certain ideas of the conventionalist critique, are nevertheless questionable in their adequacy to provide a wholly satisfactory philosophy in this context, due as we shall argue, to their misreading the role of observation in relation to theory.

Among the early, typical attempts to determine by a simple appeal to observation "which" of the metric geometries is actualized in the space of experience, measurements were made on triangles, on both a terrestrial and astronomical scale of magnitude, whose sides in each case were formed by light-rays. The basis for such attempts was the well-known differences among the three geometries relating to the angle-sum of a triangle. In the case of the Euclidean system, that angle-sum is equal to, in the Lobachevskian it is smaller than, and in the Riemannian it is, greater than two right angles. Gauss, by purely geodetic measurements, using surveying instruments, tried to measure the angle-sum of a triangle whose vertices were formed by three mountains, the Brocken, Hohehagen, and the Inselberg. The sides of this triangle were approximately 69, 85, and 197 kilometers. Within the errors of observation present, Gauss did not find any deviation from the Euclidean value. A more likely source of deviation from Euclidean values, if it was to be found at all by such direct empirical methods, was to be sought on the scale of astronomic distances. Here the efforts of Lobachevsky, Schweikart, and Schwarzschild were focussed on the problem of seeing whether one can determine a possible non-Euclideanism in the triangle formed by light-rays coming from a distant star. Where a positive parallax beyond a certain value to be found for such a distant star, it would, it was argued, be evidence for the hyperbolic character of such a triangle. On the other hand, a negative value for the parallax beyond a certain value argues for a Riemannian structure. The early efforts of Lobachevsky along these lines proved inconclusive because of faulty data. Schwarzschild, on the basis of data available to him in 1900, concluded that *if* space is hyperbolic, its radius of curvature is at least 64 light years, whereas if it is elliptical, its radius of curvature is at least 1600 light years.

Poincaré, in a frequently quoted passage, commented on efforts such as the above as follows: "If Lobachevsky's geometry is true, the parallax of a very distant star will be finite if Riemann's is true, it will be negative. These are the results which seem within the reach of experiment, and there have been hopes that astronomical observations might enable us to decide between the three geometries. But in astronomy "straight line" means simply "path of a ray of light." If therefore, negative parallaxes were found or if it were demonstrated that all parallaxes are superior to a certain limit, two courses would be open to us; we might either renounce Euclidean geometry, or else modify the laws of optics and suppose that light does not travel rigorously in a straight line. It is needless to add that all the world would regard the latter solution as the more advantageous. The Euclidean geometry has, therefore, nothing to fear from fresh experiments." [7]

The general significance of Poincaré's claims may be indicated as consisting in the fact that he brought to the fore the emphasis that needs to be given to the conventional component in all geometric description for physical purposes and the way in which this is inextricably bound up with the adoption of various *physical* assumptions. The presence of conventions means that they cannot be established as true or disconfirmed as false. Any system of thought which incorporates them, *to this extent* cannot be established as true or rejected as false. It can only be judged in terms of its usefulness and the simplicity of its conceptual means.

Conventional choices, where spatial concepts are employed in physics, occur in a number of different forms. One typical and important occasion for the entrance of a convention or stipulation occurs in connection with the adoption of measuring instruments to be used in exploring the spatial features of some physical system, for example, the shape and size of some body. This is best seen in the case of the use of some solid body selected to serve as a standard with which to measure lengths. Suppose there were some question raised as to whether the edge of a given body were *straight* or not. The normal procedure is to apply another body which is accepted as being straight to the given one to see whether the two coincide when laid edge to edge throughout their lengths to a degree of accuracy which is regarded as sufficient for the purposes at hand. Were the answer to be affirmative, the question might continue to be raised—how do we know that the body used as the standard is itself "straight?" Clearly the question can continue to be asked about any body to be introduced as in turn a "correct" measure, and so we are led to an infinite regress, that is, to the impossibility of providing a satisfactory answer at all. The reason lies in presuming that we can *know* by experience or in any other way

[7] *Science and Hypothesis*, transl. G. B. Halsted (*Foundations of Science*, Science Press, New York, 1929) 81.

whether a body *is* straight, in any ultimate or absolute sense. Actually, instead, we can only *stipulate* or *define* that some body shall be regarded as having an edge which will be *called* straight, and which will be adopted as a standard with which to judge or measure the straightness of other bodies. It is true that the question of straightness can be further analyzed, but we find that in so doing we arrive at the need to establish *another* convention in another place, so that all that is accomplished is to postpone or shift the occasion for introducing conventions rather than avoid them altogether. This is the case when, for example, we choose to characterize the straightness of some body in terms of its *rigidity*. For example, we may say that a physical line is straight when, using the rigid body as our criterion, one on which suitable units have been established, e.g., centimeters, inches or the like, and measuring the lengths of all lines joining the end-points of the line to be measured, we discover that the latter is the *shortest* of all such lines. But now the obvious question arises: under what conditions is a body *rigid*? Once more analysis shows that all that can be done is to *say* that the interval between two point marks on some selected body shall be regarded as invariant or constant. Here experience and an accumulated body of knowledge may come to guide us in the selection of the type of body to be used as our standard. Knowing the way in which various bodies expand or contract in different degrees under the influence of such factors as temperature changes, or the action of forces of a mechanical, electrical, or magnetic kind, we shall select in practice some body whose variation in these respects is comparatively small. (We may in *theory* conceive, that is, define an ideally or absolutely rigid body as one upon which no differential forces are acting. This is what Reichenbach does.[8] But it should be noted that we cannot use as a measuring instrument any such purely theoretically defined "body." One can only use some actual body given in experience and one selects it because it approximates as close as possible within practical limits to the ideal as defined.) In any case, even with such an actual body, the use of it as a standard of (relative) rigidity or self-congruence is not something which can be established simply by experience. We say it is practically rigid because it is taken as satisfying, to a degree which is sufficient for

[8] Reichenbach, Hans, *Philosophie der Raum-Zeit Lehre* (Walter de Gruyter, Berlin, 1928), 29 ff.

the purpose at hand, theoretically defined conditions. These conditions, however, involve the use of such concepts as temperature or force and the systems of thermodynamics, mechanics, or electro-magnetism in which these figure. Suppose, now, therefore, we should choose to employ some *other* body than the one "recommended" in terms of the given body of physical knowledge. For example, we may take some piece of iron which ordinarily, we say, expands under certain changes of temperature; let us define *it* as rigid even in the presence of such temperature changes. Is there any way in which this can be disproved? Again the answer must be in the negative. For the adoption of such a measuring instrument as rigid will, to be sure, require extensive changes in the formulation of the rest of our body of physical knowledge. But in making such changes we shall nowhere be led necessarily into any self-contradiction. Rather what will undoubtedly result will be an enormously more complicated system than the present one which will prove to be highly unwieldy in our interpretation and organization of the facts of experience. This consequence points the important moral: what will determine our choice of a particular measuring standard is the economy and simplicity of the *entire* body of conceptual tools for representing our experience and not simply any isolated part of it.

Another form in which the opportunity exists in physical inquiry for making a conventional decision involving the use of spatial concepts occurs in connection with the choice of one of the particular forms of metrical geometry. Just as on another level, one has, under certain conditions, the option, for example, *within* the framework of Euclidean geometry to employ one or another of several types of coordinate systems (rectangular, oblique, polar) to serve as a device of representation and calculation, so, too, on a more general level there exists the option, under certain conditions, for adopting one of the metrical geometries, Euclidean, hyperbolic, or spherical, by which to aid in the representation of the observational materials in some physical domain and to constitute part of the techniques of inference applied to such materials. Let us assume we are confronted with the outcome of measurements on the basis of (conventionally chosen) "rigid" rods, or optical devices like theodolites, or as in astronomy, photometric techniques, that some body or physical system has a particular metrical structure. For example, one may measure the angle sum of a triangle as it exists on some surface by one of these means and find that it is greater than two right angles, indicating—if the measurements are sufficiently accurate—the presence of a spherical metrical structure. Is such a disclosure to be taken as necessarily final, as the only possible way in which the metric of the surface must be described? The answer is negative, for it is always possible to elect to use another metrical scheme in calculations about such a surface. But in order to do so it will generally be necessary to reject both the original measuring instruments as no longer satisfactory and in addition it will be necessary to make other changes in the conceptual and theoretical background on a physical level that had been used to interpret the observational data. Such changes, for example, in the scheme of mechanics, or optics, may once more, when carried through, introduce a very great degree of complexity as compared with the system as it existed in its original form. Again, while such complications will be, methodologically regarded, extremely unwelcome, they cannot be taken as a sign of either self-contradiction or of falsehood.[9]

It is at this point that Poincaré's claims show a certain deficiency when looked at in a historical context. For while Poincaré argued that *in principle* one always has the option of choosing one's system of geometry, even at the cost of having to modify the rest of one's physical system, he predicted that *in fact* physicists will always choose to employ Euclidean geometry because as a matter of alleged mathematical fact it is the simplest. On both questions of fact it turns out that he was wrong. Aside from the unquestioned greater familiarity of Euclidean geometry which makes it psychologically simpler than the others, there is a serious question as to its greater logical simplicity, as recent investigations seem to indicate.[10] Secondly, the development of the

[9] Cf. Carnap, Rudolf, *Der Raum* (Kant-Studien, No. 56, 1922) 32 ff.
[10] "The recent development of hyperbolic geometry indicates that Euclidean geometry lacks even the distinction of logical simplicity. For in Euclidean geometry 'congruent' and 'perpendicular' cannot be defined in terms of intersectional constructions, and hence these (or similarly complicated) relations must be incorporated in the list of undefined concepts on which Euclidean geometry is based. Even 'between,' while capable of an intersectional definition, cannot be proved to enjoy the properties of linear order except on the basis of complicated assumptions. The only geometry which indeed is simple in the indicated sense is (as has been shown in the Reports of a Mathematical Colloquium, 2nd series) the hyperbolic geometry initiated by Bolyai and Lobachevsky. Hyperbolic geometry is the only one which can be developed from a few simple assumptions concerning 'joining,' 'intersecting,' and 'continuity' alone." Karl Menger, "Theory of Relativity and Geometry," *Albert Einstein* (*Library of Living Philosophers*, vol. 7), 464.

theory of general relativity has amply demonstrated the way
in which a successful physical theory can be set up in which
non-Euclidean geometry can be made to function as an integral element in the interpretation of phenomena. The theory
achieves its greater powers of logical unification by showing
how in place of the independence, as hitherto conceived, of
the idea of a universal force like gravitation (universal in the
sense that it acts in the same way upon all bodies and without the possibility of screening it off) and the idea of curvature of space, these may now be identified. As expressed in
the field equations and the principle of equivalence, one may
now interpret certain quantities occurring in the equations
(the "g's"), dependent upon the distribution of matter and
energy in the field, *either* as potentials and so "forces" in the
field *or* as functions of the coordinates that determine its
metric. For example, instead of saying that Euclidean geometry must be used to describe the paths of light rays and
material particles in a gravitational field, where it would become necessary to invoke the presence of universal forces to
account for the curving of such paths, one adopts a non-
Euclidean measure of straightness, i.e., geodesic paths in a
continuum with non-zero curvature.

Whatever the eventual fate of the theory of relativity may
be, the logical point it helps to establish is clear. The criterion
of simplicity which operates in the selection of theories is not
one which is restricted to the geometry alone as this is made
use of in a given theory. It is rather to the effectiveness and
logical simplicity of the theory as a whole that we must look.
That some theory, as in relativity, may achieve such simplicity through the use of a non-Euclidean geometry, or that some
other may achieve it through using a Euclidean system is
something which has to be judged in each case by examining
the theory as an integrated system of rules and equations and
not simply by judging the geometrical rules and conventions
by themselves.

The foregoing helps us to meet an argument offered by
Reichenbach that there is an empirically discoverable "natural" or "normal" geometry of space.[11] These terms are re-
served by him for that geometry employed in the description
of physical phenomena which does not require any appeal to
"universal" forces. Now actually there is nothing in principle

[11] Cf. Reichenbach, Hans, *The Rise of Scientific Philosophy* (Univ. of
Calif. Press, 1951) 1937.

which says that we might not have a physical theory in which we should choose to represent observed phenomena with the idea of universal forces rather than in terms of geometrical notions. To speak of "natural" geometry as one in which such forces are zero is to give a special prominence to one mode of representation. It is no more "natural" than any other, in the sense that "Nature" favors it because she embodies it. However powerful the geometrizing approach of general relativity is, it is this *specific theory* to which Reichenbach appeals for his *general methodological claims,* yet that theory remains after all one device among others.

4. The Metric of the Universe as a Whole

The results we have reached so far may now be applied in connection with the use of spatial concepts in cosmology as one of the specific areas in which the problems of "physical geometry" arise.

We shall consider as a typical attempt at assigning spatial properties to the universe as a whole, that which is made on the basis of the theory of general relativity with its field equations as the general theoretical schema serving as the point of departure. It will be assumed, accordingly, that the curvature of space is "determined" by the distribution of matter. In the case of the universe as a whole the matter is taken as smoothed out to yield a homogeneous and isotropic distribution. The space or metric of the universe will, therefore, be one of the three spaces of constant curvature, where the value "K" of such curvature may be zero, positive, or negative. While the type of curvature may be assumed to be constant, the theory permits the conception of a variation with "time" of the degree of curvature; it is on this basis that the notion of an expanding, contracting or oscillating space is developed. To determine what type of curvature is to be assigned to the universe two initially different procedures have in general been followed in relativistic cosmologies. These correspond broadly to the two cases discussed above in connection with the adoption of some measuring standard, on the one hand, or metric scheme, on the other, in physical geometry generally. One procedure consists in choosing at the outset one of the three metrics on whatever conceptual grounds are considered adequate. With this specification of the metric or "line-element," together with such other specifications as are provided on the basis of theory and empirical evidence, calculations are made, for example, with respect to the average density of matter in the model or other "empirically determinable" features, the predicted or derived results of which may then be compared with available data. Where such a model as a whole is empirically confirmed, it may be taken as providing an empirical check at the same time for the geometry chosen initially and used in the construction of the model. A second procedure seeks the determination of the curvature for a suitable model by examining relevant empirical evidence, for example, nebular counts, instead of making the choice initially on purely conceptual grounds. The first method is sometimes referred to as the *a priori* method, the second as the empirical one. While suggestive in some respects, this terminology is apt to be misleading. For while in one case the metric is specified in advance and in the other is calculated on the basis of available data, yet in *both* cases the setting up of a complete model which includes a metric, and which functions in turn as an inference guide in prediction and explanation of specific empirical phenomena, itself can be arrived at only by making some decisions which are of a purely conventional character while others are suggested by available observational evidence. It would be an error, therefore, to imagine that the setting up of a model is a matter either wholly of purely conventional choice or of empirical evidence. While the various components of a model, will, for different models, be chosen on varying conceptual or empirical grounds, no model as a whole is purely of one type rather than another. What may be an *"a priori"* choice for one, e.g., curvature, may, with suitably expressed qualifications, be said to be "empirically" determined by another. But the first will correspondingly have to appeal to some empirical data and the second will have to invoke the use of some conceptual and conventional materials which are not "forced" by the facts.

Let us consider by way of example and in some greater detail a specific instance of what is taken as an "empirical" approach to the determination of the type of curvature of the universe as a whole. There are several routes which may be followed here, one of which involves the appeal to data on nebular counts.[12] We first recall some basic facts concerning matters of a purely geometric character that distinguish

[12] Cf. H. P. Robertson "Geometry as a Branch of Physics" in *Albert Einstein* (Library of Living Philosophers) 313 ff.

the several metrical geometries from one another. Just as
there is a difference in the angle sum of triangles in these different spaces when we confine ourselves to spaces of two
dimensions or surfaces, so, in considering the differences of
the intrinsic curvature of each type of space in a higher number of dimensions, analogous differences occur with respect
to the metric relations in the configurations occurring in
them. Thus, in the spaces of three dimensions, the differences
among the metrical geometries may be illustrated by reference
to the formulae giving the surface area and volume of a
sphere in such spaces. For a relatively small sphere of radius
"r" the formula for the surface area is given by (1) $S = 4\pi r^2 (1 - Kr^2/3 + \ldots)$ and that for its volume by $V = \frac{4}{3}\pi r^3 (1 - \frac{Kr^2}{5} + \ldots)$, where the real number "K"
represents the curvature of space, whose value is zero if the
space is Euclidean, negative if the space is hyperbolic, and
positive for spherical and elliptic spaces (the latter being
species of a closed Riemannian space that differ only topo-
logically). These quantities, therefore, for the surface area
and the volume of spheres will differ from the corresponding
ones in Euclidean space by amounts which depend on the
value of "K"—being less than the Euclidean value if "K" is
positive and greater if "K" is negative.

Now let us return to the physical situation where we wish
to determine "empirically" the specific type of curvature to
be used in an account of the universe. Assuming the approximately uniform distribution of nebulae, the number of nebulae within a sphere of radius "r" will be proportional to its
volume "V." If we can obtain reliable observational data on
the number of nebulae out to some distance determined by
the range of our telescopes, then the way in which the number count of such nebulae varies with the distance plumbed
should provide information about the extent to which the volume explored adheres to, or departs from, the Euclidean
value for such volume. The details are given in the following
brief account by Robertson, whose discussion of this method
has been followed here. "*If* all the nebulae were of the same
intrinsic brightness, then their apparent brightness as observed from the Earth should be an indication of their distance from us; we must therefore examine the exact relation
to be expected between apparent brightness and the abstract

distance r. Now it is the practice of astronomers to assume that brightness falls off inversely with the square of the "distance" of the object—as it would do in Euclidean space, if there were no absorption, scattering, and the like. We must therefore examine the relation between the astronomer's "distance" d, as inferred from apparent brightness, and the distance r which appears as an element of the geometry. It is clear that *all* the light which is radiated at a given moment from the nebula will, after it has travelled a distance r, lie on the surface of a sphere whose area S is given by the first of the formulae [given above]. And since the practical procedure involved in determining d is equivalent to assuming that all this light lies on the surface of a Euclidean sphere of radius d, it follows immediately that the relationship between the "distance" d used in practice and the distance r dealt with in the geometry is given by the equation $4\pi d^2 = S = 4\pi r^2 (1 - Kr^2/3 + \ldots)$; whence, to our approximation [2] $d = r (1 - Kr^2/6 + \ldots)$, or $r = d (1 + Kd^2/6 + \ldots)$. But the astronomical data give the number N of nebulae counted out to a given inferred "distance" d, and in order to determine the curvature from them we must express N, or equivalently V, to which it is assumed proportional, in terms of d. One easily finds from the second of the formulae (given above [1]) and the formula just derived [2] that, again to the approximation here adopted [3] $V = \dfrac{4}{3}\pi d^3 (1 + 3/10 Kd^2 + \ldots)$. And now on plotting N against inferred "distance" d and comparing this empirical plot with the formula [3], it should be possible operationally to determine the "curvature" K." [13]

Suppose we emerge as the result of all this with some definite indication as to which of the three curvatures is suggested by this procedure. What can we say that we have established? To say that we have established the type of metrical structure of the universe is at best only a circumlocution, and if taken literally, that is, as claiming to give us knowledge determined by experience of what holds in fact, antecedently or independently of our inquiry, is surely not warranted by any inspection of what the procedure itself exhibits. We do not discover the metric of space along with or in the same way as we discover the numbers of nebulae of given photo-

[13] Robertson, H. P., *loc. cit.*, 330 f.

graphic apparent magnitudes. The latter can be represented by the former, but whether we represent that number by a particular metric i.e., by the use of a particular geometry is altogether a matter of choice, or if one prefers, of a discovery of another kind, not one of observable fact but of representational efficacy. Appeal to nebular counts involves conventional decisions just as much as the use of measuring rods or light rays. Just as in the case of measuring rods we posit their congruence and rigidity under transport, and with light rays employ a mode of representation according to which, for example, they are said to travel in straight lines (which is by no means found *in fact*, but rather adopted because of its overwhelming usefulness in geometric and physical optics), so too, in the case of nebular counts, we posit or presume that the data obtained from distant nebulae can be interpreted by appeal to the same laws or theories as are used in other and "nearer" domains. This is the case, for example, when use is made of the inverse square law of light diminution for determining the "luminosity-distance" of nebulae and the distinction between their apparent and absolute magnitudes. The same type of decisional element is involved in the interpretation of the spectroscopic data, for example, in connection with the red-shift phenomena, wherever one presumes that the same laws or theories are to apply to the analysis of radiation arriving from nebulae as from any other source. Or again, there is a convention involved in assuming that all the nebulae, at whatever distance, have the same average intrinsic luminosity. The decision to use such common principles of interpretation is, to be sure, guided by considerations of simplicity. Why appeal to unknown principles if it is possible to get along with ones already used successfully elsewhere? But once more the criterion of simplicity is a relative matter and the use of such common principles is by no means forced by the facts themselves. Nor does it do any good to appeal to the so-called Principle of the Uniformity of Nature. This, at best, is a disguised way of expressing the same regulative canon of using the same formulae wherever possible; it is not, however, a binding constitutive "truth" of either an axiomatic or synthetic nature.

There is, then, no simple, exclusive coordination of metric scheme with nebular counts taken by themselves. Nebular counts are nebular counts and never, as such, require—no matter how extensive the surveys—that we use a particular metric scheme to describe them. Rather there are a great number of observational facts of a type logically on a par with nebular counts, e.g., red-shift and magnitude data. Such a group of facts is never closed either with respect to type of included material, amount, or accuracy, and needs to be interpreted by theory. The latter in turn is a coordinated set of rules of inference and representational devices, including metric ones. It is the function of the theoretical set as a whole to represent the observational data as a whole. The adequacy of the metric, therefore, cannot be judged simply by reference to one or more parts of the observational materials. It is the entire theory which has so to be judged. To speak of the metric properties of the universe or the curvature of space as a whole, therefore, is at best to refer to *one* component of a total cosmological theory. It does not refer to anything which by itself can be disclosed in observational experience. It may be more convenient and fruitful to use one metric scheme rather than another, in conjunction with other physical ideas, but we cannot say that the facts force this upon us. The problem is: does the adoption of a particular value for the curvature enable us, when taken in conjunction with the rest of some theory to make correct inferences *from* given facts *to* other facts? To the extent that the accuracy and range of such inferred observable facts is greater as accomplished by one theory with its associated metric as compared with others, to this extent we may say "the metric of the universe is such and such." The latter expression, however, is nothing more than a telescoped manner of referring to the relative superiority of the given *theory* or model of the universe.

Chapter 8

The Age of the Universe

THE DIFFERENCES that exist in current scientific cosmology on the question of the "age of the universe" are not solely ones of technical decision as to the kind, amount, or consistency of the evidence available for the assignment of a quantitative estimate. They are also, and more importantly, of a type which reflect basic disagreements of physical theory and philosophic outlook. On the level of physical theory some views—the "evolutionary" theories as they have been called —undertake to assign a finite age to the universe, while others, of which the "stationary" model of continuous creation is a prominent current example, consider the concepts of "origin" and "age" as inapplicable to the universe and insist on their being replaced by the use of an "infinite time-scale." The philosophic differences which emerge in such discussions are of an even more fundamental character. They result from failure to agree upon the answer to be given to the root question: What does it *mean* to assign an "age" to the universe, or conversely, to refuse to assign such an age, to speak of it as being of "infinite duration"? By "meaning" here is to be understood the question of the cognitive status possessed by whatever answer is preferred, not simply one of determining the technical definitions given to these terms in the respective theories.

Is it the case, for example, that acceptance of an "evolutionary" theory, one which assigns an "origin" and "age" to the universe, implies a belief in an "absolute Beginning" or "Creation"? Many, including some authors of such theories themselves, answer in the affirmative. Milne, for example, has argued for the identification of the singular event, referred to in his theory as occurring at $t = O$, with the idea of Creation as this is understood in one form of traditional theology. Many who are not necessarily prepared to accept Milne's own cosmology, have followed him in this type of appeal by arguing that other scientific cosmologies, for example, some relativistic ones, that speak of a "beginning" (of expansion) are also supports for the belief in an absolute origin. While surely not all cosmologists, even among those defending an "evolutionary" scheme are found to take such a view with respect to the notion of an age for the universe, some indeed deliberately rejecting any such associations, yet disagreements of a philosophic or methodologic nature continue to manifest themselves with respect to the question as to whether and in what sense, theologic associations apart, it is at all possible for scientific cosmology to seek for an empirical determination of the age of the universe. Since the origin of the universe, if such there was, is no longer observable, the question is whether there are any facts to be established by observational means *now* which would suffice as a basis for reliable inference as to when such an origin did occur. There are some who would seem to argue that despite the admittedly incomplete and relatively inaccurate observational evidence at present available, such a determination is possible *in principle*. What we must ask ourselves, accordingly, are these questions: In what sense might scientific cosmology be able to determine whether the universe had a beginning or not (and so a finite age or not)? Could it in any case make such determination on empirical grounds?

It will be one of the purposes of the following discussion to show that not only is the familiar and unfortunate practice of interpreting the outcomes of such evolutionary cosmologies in terms that are borrowed from one form of traditional theology or other speculative schemes wholly unjustified, but that even any supposedly purely "scientific" claim to be able to establish an origin for the universe on empirical grounds, in the way in which this is frequently understood, is similarly a wholly unwarranted gloss on what such theories in their actual functioning are at their best able to accomplish. It seems like a harmless and natural transition to go from speaking of "John Smith" or "the Earth" to "the universe" as if each was an individual in the same way and therefore could sustain predicates in the same ways. Thus we might speak clearly and unequivocally of the "age of John Smith" and we could presumably speak in the same way of "the age of the Earth." Why should we not be able to speak similarly and without any special complications or obscurities of "the age of the universe"? To think so is what underlies a good deal of the confusion engendered in cosmological discussion. It is particularly encouraged by the uncritical habit of extending the grammar of conceptualizations from one case to another without stopping to ask whether there are significant differences which stand in the way of such extrapolation or extensions. The interest of the following analysis is not to criticize the use as such of evolutionary theories nor to favor in principle those theories which make use of an "infinite time-scale," since logical difficulties of a parallel sort are frequently noticeable in those accounts which make use of the latter concept as well. A cosmological theory which proposes to account for the various features of the universe already observed and which might serve as an instrument for further predictions, *may be* of an evolutionary type. As such it will be founded on whatever usefulness there is in regarding the universe as a developing entity with an "origin" and an unfolding history. Nothing can be set up *a priori* as a permanent block against the use of such a theory, nor can we overlook the possibly great advantages offered by this type of model. But it will be argued that precisely because any theory whatsoever, and therefore such theories in particular, are what they are—devices of representation and calculation—one must reject any interpretations put on them which takes them for something else, as giving us a kind of *information* or knowledge which is not in fact obtainable from them. The fact is that the concepts "origin" and "age" when used in conjunction with "the universe," and understood in terms of the way in which they function in the context of cosmological theory, must be recognized to have a different status from the way in which these same terms are understood when used in reference to objects or processes of a familiar empirical sort.

1. Time and the Use of Clocks in Physics

It will be well to begin with some comments on the way in which time-measures and the use of temporal predicates make their appearance in physics generally, before turning to the specific problems that confront us in the field of cosmology.

Our interest in the measurement of time is summed up by the desire to receive answers to the questions "how long?" and "when?" An answer to the first type of question provides us with a means of quantifying the *duration* of events and processes, while the second provides us with a means of indicating the *position* of an event in a sequence of events with respect to the ordering relation "before and after." A *clock* is any device or process which is selected for the measurement of the *duration* of intervals. The setting up of a system for the *sequential ordering* of events, that is, a device for the placement of any event with respect to its "before and after" relations to other events is accomplished by means of a temporal co-ordinate frame of which the *calendar* is a simple and familiar example. To "tell the time" by clock or calendar, as these are ordinarily employed, requires, moreover, if we would not talk nonsense in our telling that we be able to identify directly or by inference the processes or events whose sequences are to be described or whose durations are to be measured. To construct a "calendar" or temporal co-ordinate frame requires that at least two conditions be fulfilled: we must select some event which will serve as our "origin of co-ordinates" (such as the birth of Christ, or "now" for some particular observer, or the like) and we must be able to "date" all other events with respect to this origin by noting its position, that is, whether it occurred earlier, at the same time, or later than the given event. In order to accomplish such "dating" of events in a standard way, and, in any case, to be able to say "how long" events endure, even apart from dating them with respect to some chosen origin of co-ordinates, resort is made to clocks. What are called "clocks" in the popular sense, e.g., the timepiece which is operated by a spring, pendulum, or other devices, are in fact mechanisms with a double function. They not only measure durations (as, for example, "stop-watches" are designed especially to do) but they serve a "dating" or "calendar" function as well. Thus the ordinary "clock" tells us not only how long, for example, an hour is, it also tells us that something happened, let us say at "five-o'clock." The latter is a "date" with respect to noon or midnight, the two "origins of coordinates" that serve in each day for the two twelve hour periods. It is with clocks functioning simply as instruments for measuring the *duration* of intervals that the *additive* properties of time and all the mathematical operations based on this become significant.

In order to make use of such additive properties, however, some method must be instituted for specifying what is to be meant by "equal" times. The conditions to be satisfied in determining the equality of events as measured by clocks recall methodological considerations of a type similar to those encountered in connection with the measurement of spatial intervals (lengths). On a first level of approximation, that is, even before explicit use is made of theories as guides, clocks are set up wherever, as normally is the case, some cyclic, repetitive process is chosen whose periods are easily demarcated and can be used as standards for comparison with the duration of other events. Such standard intervals are found in the diurnal rotation of the earth, pulse beats, the emptying of a fixed amount of sand through an orifice, the swing of a pendulum of fixed length and similar natural or man-made cyclic or repetitive processes. Suppose, in taking any one of these actual "operationally defined" clocks as our standard for the measurement of other events, a question is raised as to whether the successive periods of the clock itself are equal or congruent with one another. Is there any way in which we can establish such equality or congruence of successive intervals in a given clock? Since the passage of one such interval cannot, as a passage, be preserved for comparison with a succeeding passage of another interval of the same clock, how can we ever *know,* it may be asked, whether the two are indeed equal? It might be thought that we can know this by reducing the measure of time-intervals to some other measure, for example, space. But this will not do. For even if one were to confirm that, for example, equal distances are traversed by some moving body taken as a clock, the question is whether the *rate* at which such distances were traversed was the same in all instances, and so we are led ultimately to recognize the irreducibility, as a matter of knowledge, of time measures to anything else. We may *define,* if we wish to, although there is no necessity to do so, clock measures in terms of e.g., distances traversed, but this is not a question of *knowing* the equality of time intervals thereby. One might, on the other hand, undertake to compare the beats or phases of one clock with that of another clock, for example, the beats of the pulse with the swing of the pendulum, or the solar day with a sidereal day. But whatever the outcome, that is whether we find the intervals to agree or not, the same question obviously can continue to be raised about whatever clock we appeal to—how can we know whether *its* successive periods are congruent? Once more, as in the case of the measurement of spatial intervals of length, the answer must be that any determination of congruence of time-intervals is at some point not a matter of knowledge at all, but for *definition.* It becomes a matter of convention or stipulation; we *call* the successive intervals of some selected clock *equal.* The only significant question is not whether this is true or false in a given case, but whether it is convenient for a given branch or *system* of knowledge to do so. It is at this point, in the measurement of time as in the measurement of space, in chronometry as in geometry, that it becomes important to recognize that the instruments used and the methods of representation employed for codifying their results or for performing calculations are not isolated or self-contained topics. Conventional choices, though free, need not be arbitrary. While, strictly speaking, the use of any given clock whose periods are defined as constant, need not lead to any contradiction in the system of concepts that incorporates its use, provided adjustments are made in the entire system of such concepts, in practice of course, the cost in complications resulting from such adjustments in order to adhere to certain clocks, recommends for such cases their abandonment. The system of knowledge as a whole becomes more effective, because logically simpler, with the choice of a new definition of "equal times."

To provide such a definition, the conceptual resources of physical theory constitute an important level of advance in the use of the notion of time. As distinguished from the "operational" appeal to actual instrumental devices or observed natural processes, theory now makes available, among other things, the notion of an *ideal clock*. While the materials for setting up such an ideal may be suggested by observed processes or even by constructed devices, the "ideal clock" cannot be identified with any of these; indeed on the basis of the ideal it becomes possible to specify the limitations and inaccuracies of actual clocks and to provide corrections for them. However, in speaking of an "ideal clock" it must be said that this is actually a compressed way of saying that one is going to appeal to a certain *formula* in one's calculations. The way in which the "intervals" of such an "ideal clock" are defined as equal is accomplished by the adoption of a principle of some theory to serve as the definition of the mechanism of such a clock. Since the statement of this principle, and the clock which it helps define, is normally part of a wider and more complex system of conceptual relations making up a total physical theory, the question of the "correctness" of the clock reduces to one of the effectiveness in which the theory functions as an integrated whole. These remarks must now be clarified and we may do so in terms of some familiar instances. That, for example, the rotating earth is not an adequate clock for purposes of very refined measurements or calculations because of the effect of tidal friction, (as well as other forces) in slowing down its rate of rotation, came to be realized with the adoption by the theory of Newtonian mechanics of a standard which involved, ultimately, appeal to the principle of inertia. By the aid of this principle it is possible to define an ideal clock as one to be identified with the motion of an absolutely free body, that is, a body not acted on by any external forces. For "equal times" means "the intervals elapsed in the motion of a free body as it traverses equal distances." It is clear, at this point, that what is at stake in any questioning of the "correctness" of a freely moving body as an "ideal clock" is no longer an isolated matter of time measurement but involves a complex system of rules and conceptual devices which can claim in its behalf a varied and extensive body of successes in explanation and prediction. Among other things, the advent of the special theory of relativity, in challenging the adequacy of the classical system, showed how, in using the Lorentz equations, crucial aspects of the older definition of time had to be surrendered. On the view of special relativity, for example, two events judged to be simultaneous by an observer attached to one inertial frame will not be so judged by an observer moving with respect to the first with some uniform velocity. There will be similar differences in their respective measures of the duration of some independent event, At bottom these differences between special relativity and classical mechanics (differences that only become significant where very large relative velocities come into play) are based on the revised conception inherent in special relativity of what an ideal clock is. Such a clock may be defined by using for this purpose the principle of the constancy of the velocity of light, one of the fundamental bases of the special theory. This principle asserts that regardless of the relative motion of the source of light or the observer, the velocity of light will have the same constant value for all observers in uniform motion with respect to one another. With the aid of this principle, a moving light ray or wave front may be set up as an "ideal clock" since the "equal times" it "beats out" are defined as those to be measured by the equal distances it traverses. But why should the conception of clock measures introduced by special relativity be preferred to the classic one? Again this is not a question to be answered by considering clock measures alone; it is rather a question of the effective superiority of the newer theory as a whole over the older one, a superiority which carries along with it its altered view of time relations. In no case can we speak of approximating to an absolutely perfect clock, for this would mean that we can also speak of an absolutely perfect theory, and this is an idea which has no methodological significance.

2. "The Beginning" of the Universe

Let us now return to the field of cosmology and ask in what sense one might speak of "an age of the universe." Since the determination of an age requires some means of measuring or calculating the passage of time from a recognized origin, what kinds of clocks or temporal coordinate systems are to be used in connection with the determination of an "age" for the universe, and how are the results of such measurements and calculations to be interpreted?

The pattern of one typical answer, as given by those who adopt an "empirical" approach in support of an evolutionary cosmology, consists in appealing to certain allegedly relevant natural processes about which reliable observational evidence is to be had. These processes are said to provide natural "clocks" whose rates (being assumed constant) enable us not only to measure "equal" time-intervals, but when used to set up a temporal co-ordinate frame, sanction inferences to those events said to mark the "origin" of such processes themselves. It is the approximate agreement of the use of several such natural "clocks" as they converge on a point occurring several billion years ago (estimates vary, but cluster, in present discussions, around values from $3-6 \times 10^9$ years) which is regarded as a natural "origin" for the universe and its age reckoned from that "origin." The type of evidence characteristically appealed to here includes, in one direction, age-estimates obtainable for such astronomical bodies or systems as the Earth, individual stars, star clusters, and our own Galaxy, that are taken to constitute at least a lower-bound for the age of the universe as a whole. These estimates are based on the use of such "clocks" as radio-active decay rates for judging the time required to produce radiogenic lead in rock-ores on the surface of the Earth, the rates of thermonuclear conversion of hydrogen into helium as a basis for estimating the evolutionary development of stars, the establishment of dynamical stability for dense or loose star clusters and binary systems. Again, for "the universe" proper, the Hubble-Humason formula for the red-shift in the spectra of nebulae is taken as a sign of their mutual recession and thus evidence for the expansion of the entire system from an initially highly condensed state, which is regarded as the "origin" of the system as a whole.[1] Recent revisions of the scale for intergalactic distances (accomplished by Baade and others) as incorporated in the Hubble formula, give a revised "time-scale," which, being at least twice the former values, removes certain earlier discrepancies with geophysical and other data, and establishes, according to this view, an "age" of the universe of the magnitude mentioned (roughly 5×10^9 years).

With respect to this type of approach to the question of "the age of the universe" a number of critical comments are in order.

(1) In the first place, it needs to be said that however striking be the agreement among the age-estimates for some of the astronomical bodies indicated, planets, stars, or individual galaxies (and even were there no single figure arrived at of the same rough order of magnitude) such age-estimates do not as such *require* that they be at the same time interpreted as making sense *only* on the basis of an evolutionary *cosmology*. The age-estimates of these various bodies rest upon the use of "clocks" and theories that are applied specifically to the account of the origin and development of these several types of bodies. There need be, indeed there is no serious question raised about the aging of these entities. Yet one might give an acceptably reliable account of the origin and development of planets, stars, and individual galaxies without being compelled on logical grounds to accept an evolutionary approach or theory that would assign an origin and age to the universe as a whole. One might go even further. We might recognize aging not only in such empirically identified objects as planets, stars, or individual galaxies, and even be prepared to affirm such an evolutionary development in the *population* of galaxies at present under observation, without being logically compelled to accept an evolutionary cosmology which would ascribe an "origin" and "age" to the universe-as-a-whole. In giving an account of the universe-as-a-whole we are, at best, making use of a *theory* for the interpretation of the data of observation. Possibly successful theories of the universe-as-a-whole (the continuous creation theory is one such current candidate) might be constructed which, while admitting indeed the existence of evolutionary

[1] Cf. Bok, B. J., "The Time-scale of the Universe" *Mon. Not. Roy. Astr. Soc.*, 106 (1946) 61; Tolman, R. C., "The Age of the Universe," *Rev. Mod. Phys.*, 21 (1949) 374; Bondi, H., *Cosmology*, Ch. VI; Gamow, G., *The Creation of the Universe* (New York, 1952); Opik, E. J., "The Age of the Universe," *Brit. Journ. Phil. Sci.* V, 1954, 203.

development in various astronomical subsystems and even in the population of the observable region of the nebulae, might nevertheless uphold a conception of the universe as a whole to which *no* evolutionary development is to be assigned.

(2) With respect to the facts connected with the universe of galaxies which are claimed in support of an evolutionary cosmology, it is necessary to insist that the use of Hubble's law of red-shifts as providing a "clock" for determining the age of the universe, an age to be reckoned from the origin identified with the initially high condensed state marking the beginning of the expansion of the system of galaxies, represents one, but by no means the only way, of regarding that formula. Hubble's law summarizes the relationships between the result of measurements of the line-displacements of the spectra of nebulae and the apparent magnitudes of such nebulae. It is an empirical generalization to the extent that it connects these particular quantities; it does not demand one rather than another interpretation or correlation with the conceptual constructs of some particular theory in order for its own intrinsic or observationally-relevant meaning to be understood or its validity established. To regard its ulterior significance as marking an actual recession of the galaxies or as having some altogether different "meaning" is a matter of *theory*. The judgment as to which particular theoretical interpretation will be most adequate is to be determined by reference to the efficacy of the entire theory of which it forms a part. Evolutionary theories do not necessarily enjoy a permanent guarantee of success here.

(3) Inasmuch as Hubble's law does not by itself constitute a clock for measuring the age of the universe, but can only be interpreted on the basis of some theory as doing so, it is important next to examine in what sense, when such an interpretation *is* offered—as by an "evolutionary" cosmology— one is to understand the assignment of an "origin" and an "age" to the universe. For this purpose, and by way of illustration, we may turn to the "expanding universe" models of general relativity to see what kind of methodological transformations are involved in converting a relationship like Hubble's law into a clock for reckoning the "age of the universe."

As we have seen, the general theory of relativity involves as a general representational device the adoption of a Riemannian four-dimensional space-time manifold for the mapping of events, three co-ordinates identifying the spatial location and one co-ordinate the temporal location of any event. The motions of free particles and light rays are represented by geodesics in such space-time. The particular metric correlated with a given distribution of matter and energy is governed by relations expressed in the field equations of Einstein's general theory of relativity. For the situation faced in cosmology, these equations reduce to two differential equations which express the relationship between a certain quantity $R(t)$, the "radius" of space at time t in terms of the density $\rho(t)$ and pressure $\rho(t)$ of matter and energy at this same time. The equations also contain the constant k for the curvature of space (whose value is zero if space is Euclidean and infinite, $+1$ if spherical and of finite volume and -1 if hyperbolic and also of infinite volume) and the cosmological constant λ. To solve these equations would mean to be able to determine the value of R for all values of t, that is, to be able to specify the entire "temporal behavior" of the model. For this purpose various clues are sought in existent empirical findings and among these the "law" of red-shifts figures prominently. In relativistic cosmology, generally, the concept of an expanding "space" is used in turn as a basis for explaining the red-shift. To put the matter roughly, the "radius" of "space" is considered to be a function of "time" such that the "volume" expansion of "space" is the reason for the separation of nebulae noted by their red-shifts. With space of increasing "radius" the material particles identified with the nebulae and assumed to be initially homogeneously distributed, remain scattered uniformly throughout, while their mutual distances increase in the same ratio. It is this increase in their mutual distances which appears to us in the form of their velocities of recession being proportional to their distances. A postulate due to Weyl and adopted in most schemes of relativistic cosmology permits the representation of the world-lines of nebulae as a bundle of geodesics converging to the past. With the aid of this representational device, it is possible to introduce a family of 3-spaces orthogonal to the world-lines of all nebulae, where each such 3-space identifies a hypersurface for which "t" is a constant, and in which all nebulae are at relative rest when measured by the use of co-ordinates attached to such a "surface." All such 3-spaces are taken as possessing a uniform curvature. It is then, the variation in the "radius" of "space" affecting such a curvature and changing in some determinate way with the "passage" of "cosmic time" which marks the over-all "motion," e.g., expansion of the universe as a whole. Here a certain caution and refinement in the language used to translate the mathematics is advisable. Since the schema of an expanding universe allows for a space which is either infinite (i.e., Euclidean or hyperbolic) or finite (i.e., spherical) reference to a "radius" of space undergoing "change with time" has to be understood in a generalized sense. As de Sitter puts it: "The interpretation of the expanding universe, the making of a mental picture, or a model of it, which was, or appeared to be, easy when we knew that the universe was finite, is not such a simple matter now that we do not even know whether the curvature is positive, zero, or negative, whether the universe is finite or infinite. It sounds rather strange to talk of an infinite universe still expanding. If we were certain that the curvature was negative, we might still, as in the case of positive curvature, replace the phrase "the universe expands" by the equivalent one "the curvature of the universe decreases." But if the curvature is zero, and remains zero throughout, what sort of meaning are we to attach to the "expansion"? The real meaning is, of course, that the mutual distances between the galactic systems, measured in so-called natural measure, increase proportionally to a certain quantity R appearing in the equations, and varying with the time. The interpretation of R as the "radius of curvature" of the universe, though still possible if the universe has a curvature, evidently does not go down to the fundamental meaning of it. The manner in which time and space are bound up with each other in the four-dimensional continuum is variable. It is difficult to express this variability of the cross-connections between space and time in simple language, and different interpretations of it are possible, corresponding to different mathematical transformations of the fundamental line-element, e.g., a different choice of the variable which we interpret as "time." Perhaps the best way we can express it is by saying that the solution of the field-equations of the theory of relativity shows that there is in the universe a tendency to change its scale, which at the present time results in an expansion, but may perhaps at other times become, or have been a shrinking." [2] The construction of individually distinctive models within the framework of common assumptions characterizing relativistic cosmologies, that is, the solu-

[2] de Sitter, W., *Kosmos* (Harvard, 1932) 128 f.

tions to the equations giving the complete picture of "R" as a function of "t" proceeds by substituting values for some of the quantities occurring in these equations, values thought to be reasonable and in accord with reliable empirical evidence now available, and then solving for the remaining "unknown" quantities. Differences in models, i.e., in solutions obtained and in the approach adopted toward such solutions, are bound up with the quantities selected as "known" and the reasons given for such selection. On the basis of these various assignments of values, one finds among other things, different accounts of the variation of "R" with "t." In particular, some non-static models involve the identification of a point on the temporal co-ordinate frame for the universe as a whole which is taken as marking "the beginning" of its expansion. In some cases this may be identified with "$R = O$," in others with a minimum finite value of "R." It is with respect to such models that it is customary to speak of their providing a finite "age" to the universe, an "age" that is to be reckoned from the initial singularity identified with such specific values of "R." [3]

It must be stressed in connection with the construction of these various models of the "expanding universe" that Hubble's "law" does not as such provide a unique and unequivocal basis for determining the "rate" at which expansion may be conceived as having taken place. To convert Hubble's law into a clock for calculating the age of the universe requires the selection of one of several open possibilities allowed for in the schema of relativistic cosmology. Thus, whether the expansion is to be interpreted as a linear one or one of several nonlinear types, cannot be found by inspecting the empirical law itself, but only by making these several choices in turn, fitting them into the framework of the relevant equations and seeing which resulting solution serves as the most satisfactory basis for representing the various facts known. As McVittie points out, however, none of the various possibilities at present recognized are as yet entirely free from difficulties of a conceptual sort. For example, certain choices which at first sight seem most plausible lead to the conclusion that throughout the major portion of the "life-history"

[3] For the details see: Robertson, H. P., "Relativistic Cosmology," *Rev. Mod. Phys.*, 5 (1933) 62; Tolman, R. C., *Relativity, Thermodynamics and Cosmology* ch. X; McVittie, G. C., "The Age of the Universe in the Cosmology of General Relativity" *Astr. Jour.* 58 (1953) 134.

of the universe it is necessary to ascribe either negative pressures or negative densities to the model. He concludes "that the time that occurs in the so-called 'Hubble constant' is insufficient by itself to determine the age of the universe. An age can be calculated only by making additional unverifiable hypotheses as to the initial character and the subsequent course of the expansion, each set of assumptions leading to an age that differs widely from other, equally plausible ages." [4] In any case, whether the situation here described be a transitional one or not, it is clear at least from a logical point of view that to use Hubble's law as a "clock" is to *stipulate* on the basis of some choice made within and in terms of some *theory* that such and such is to be considered the "rate" at which one is going to calculate the "passage" of "cosmic time." The correctness of such a "clock" is once more, as in other instances of definition of clocks by theory, to be judged not in terms of any isolated criterion applied to the "clock" alone, but in terms of the success, i.e., the internal logical consistency, simplicity, and empirical corroboration achieved by the total theory in which it functions.

Suppose now, for argument's sake, that some quite satisfactory evolutionary theory were to emerge, free from serious difficulties and involving by its very nature, reference to an "origin" of the universe and an "age" calculated on the basis of this origin. What would it mean to speak of such an "origin" as one having occurred at "$t = O$"? What it would mean is that one has used a *theoretically* stipulated "clock" such as the "expansion of space" as a "process" which began with a "very small" or even "zero volume." Such an "origin" is nothing more than a conceptually identified point on a temporal coordinate frame which is reached by the use of a formula which represents such a theory-defined clock. Any attempt at finding meanings for what is thus referred to as an "event" marking the "origin" of the universe, or its "age" as the "cosmic time elapsed" since such an "origin," which fails to see in these ideas, devices of representation employed within a theory, overlooks the logical essence of the situation. No more than one finds geodesics or radii of curvature or any of the other associated techniques of interpretation and calculation among the facts of observation or any summary of them, do the ideas of "$t = O$" as the "origin" of the expansion of the universe or the notion of "cosmic time" rep-

[4] McVittie, *Astron. Journ., loc. cit.*, 134.

resent empirically identifiable facts. Such terminology must not be saddled with meanings which claims for them a factual significance, as having necessarily an observational or existential reference. If found useful at all, these ideas are bound up with an entire scheme of ideas which lends them significance and cannot be extracted from that context to be given any independent empirical meaning.

Now in saying all this there is no need to deny, of course, that one can make inferences *in accordance with* a theory employing such ideas to *actual events* in the past of the collection of nebulae. On the contrary, it is the primary value and function of such a theory to do precisely this. In doing this, one appeals to certain facts observed at present and, using the theory, makes inferences to earlier states. But the fact that, on the basis of some particular evolutionary theory, one cannot make inferences to any actual events which are earlier than the "event" *taken in the theory as origin,* does not mean that there are not or could not have been in fact such *actual events.* All it means is that the particular theory is unable to guide us in responsible inferences to such actual events. And this clearly is a relative matter. To claim, therefore, as some do, that the "event" taken as "origin" by a particular theory *forever* shields human inquiry from piercing behind its facade in trying to discover the "earlier" stages of the career of the universe is to misrepresent the entire situation. W. M. Smart, for example, writes: "There may be —and I think that there is—a point in the remote past where an 'iron wall,' of cosmic fabrication, shuts us out from scientific contemplation of the antecedent state and the primeval evolutionary history of the Universe; but, until this point is reached, scientists have a legitimate domain of exploration in which, indeed, many triumphs have already been won." [5] Such too, in effect, is the view of Gamow. He speaks of the "Creation" of the universe in terms of a singular state some five billion years ago in which the galaxies were packed closely together, before the general expansion began, in which the densities were sufficiently enormous to provide the thermonuclear conditions for the synthesis of the elements. As to the possibility that there may have been antecedent states of the universe prior to the "cataclysm," perhaps of a cyclic nature, Gamow does not fully commit himself. But he does claim that for all practical scientific purposes such a possi-

[5] Smart, W. M., *The Origin of the Earth* (Cambridge, 1951) 1.

bility is one that could *never* be verified. Such earlier states, if there were any, lie completely hidden from our view. Whatever might be the mathematical "attractiveness" of postulating an endless cycle of expansions and contractions, physically there is no warrant for such a hypothesis. For "most likely the masses of the universe were squeezed together to such an extent that any structural features which may have been existing during the 'pre-collapse era' were completely obliterated and even the atoms and their nuclei were broken up into the elementary particles (protons, neutrons and electrons) from which they were built." He would refer to such a "dim pre-galactic past" as a "metaphysical St. Augustine's era," "since it was St. Augustine of Hippo who first raised the question as to what God was doing before He made Heaven and earth." [6] Again, E. T. Whittaker writes that "different estimates converge to the conclusion that there was an epoch about 10^9 or 10^{10} years ago, on the further side of which the cosmos, if it existed at all, existed in some form totally unlike anything known to us; so that it represents the ultimate limit of science. We may perhaps refer to it as the Creation . . . The Creation itself being a unique event is of course outside science altogether." [7]

All such views commit the methodological error of indulging in an *ignorabimus* which is nothing short of a dogma. It erects a situation which is relative to the capacities of some specific theory into an absolute and permanent limitation. (But should we have said in a parallel case that Dalton's theory of the atom or Bohr's gave us the last word?) We always have the privilege of using whichever scheme of ideas seems most fruitful. And some theories, even if they are of an evolutionary type, might very well be able to map events by their representational devices which would give us "insight" over a much more inclusive "time-span" than some given theory accepted at the moment which, because of its specific intrinsic limitations, chooses to begin with the "event" *it* calls "$t = 0$."

[6] Gamow, G., *The Creation of the Universe* (Viking, New York, 1952) 29 f.
[7] *Space and Spirit* (Nelson, Edinb., 1946) 118, 121.

Chapter 9

Continuous Creation

IN THE recent, much-publicized cosmology of "continuous creation," [1] a number of issues of specifically philosophical interest are raised, whose resolution proves essential for an understanding and evaluation of the theory as a whole. These issues revolve about the precise *meaning* to be given to the concept of creation as it functions in the theory and the *methodological grounds* offered by way of justification for its use. Accordingly, whatever might be the fate of the theory as it undergoes the ordinary checks of mathematical analysis and observational verification, I shall argue that much of the philosophic (sometimes euphemistically called "aesthetic") commentary provided by its authors exhibits a number of difficulties and obscurities that stand in the way of its total acceptability as formulated at present.

1. A New Cosmology; Bondi-Gold, Hoyle

The essential novelty of the theory consists in the suggestion that matter is being created in all epochs and throughout space at a statistically uniform rate which is sufficient to compensate for the continuing expansion of the universe, and thus to maintain the universe in a steady-state (in a generalized hydrodynamic sense) and at an over-all constant density. The expansion is inferred from the observed red-shift in the spectra of the galaxies, which is interpreted as due to their recessional motion. Matter, it is proposed, is being created in an elemental form at random throughout space in a way which is independent of the matter already existent in various stages of agglomeration. The calculated rate of creation is roughly the mass-equivalent of one hydrogen atom per litre of volume every billion years. It is accordingly a process which, because of its virtually infinitesimal

---

[1] Also known technically as the "steady-state theory of the expanding universe": the original papers are: Bondi, H. and Gold, T., *Mon. Not. Roy. Astr. Soc.*, 108 (1948) 252; Hoyle, F., *Mon. Not. Roy. Astr. Soc.*, 108 (1948) 372; Hoyle, F., *Mon. Not. Roy. Astr. Soc.*, 109 (1949) 365; cf. McCrea, W. H., *Endeavour*, 9 (1950) 3; Bondi, H., Cosmology (Cambridge, 1952) ch. 12.

proportions, is incapable on practical grounds of direct confirmation. Through a gradual process of condensation and accretion, conglomerate macroscopic bodies are eventually built up to the enormous proportions of galaxies and clusters of galaxies. At any given time, within a sufficiently large volume of space, the number of galaxies remains constant since new ones are continually being formed to replace those disappearing over the "horizon" of the observable universe. The theory allows for the idea of a universe whose existence in time is infinite both in the past and the future, and whose existence in space is likewise infinite in extent though the range of observability is set by the limiting velocity of light.

2. The "Physics" of Creation

It is claimed for the conception of creation as it figures in this theory that it is wholly removed from the intellectually suspect domains of metaphysics and theology and now finally established within the domain of scientific accuracy and comprehension. On earlier theories "it was believed that the creation process occurred at a definite ascertainable epoch of the past. Properly interpreted this point of view sidesteps the whole question; for unless we say *how* the creation occurred, nothing has been achieved. Indeed the word "creation" in this context was simply a device for terminating the discussion as soon as an awkward question had been posed." [2] With the steady-state theory "the problem of the origin of the universe, that is, the problem of creation, is brought within the scope of physical inquiry, and is examined in detail instead of, as in other theories, being handed over to metaphysics." [3]

With respect to these claims, however, it is important to keep two points clearly distinguished. One has to do with the consequences or effects of matter when "given" as existing, the other has to do with the possible antecedents or causes which bring it into existence. In discussing creation, the authors of the steady-state theory believe it sufficient to explore the former without any attention to the latter. It is here that there is to be found one crucial philosophical difficulty in their account. Thus one set of questions has to do with the specific properties which matter is said to possess when it appears or, allegedly, is created: properties such as

[2] Hoyle, F., *New York Times Magazine*, June 1, 1952, 12.
[3] Bondi, H., *Cosmology*, 140.

spatial distribution, rate of appearance, initial velocity, "temperature," atomic structure and the like. This is what is referred to as the "physics" of creation. It is the filling out of these details, the drawing of their consequences, i.e., the linking of them with the facts of astrophysics, atomic theory, and observational astronomy, which underlies the claim that creation is therewith brought within the scope of scientific understanding. And there can be no doubt that progress in this direction makes for the kind of deepening and broadening of insight which is characteristic of science. *Yet the possible success of this line of inquiry would leave completely unaffected the crucial claim that what has been investigated is matter which has been created*. Those theories like Lemaître's or Milne's which posit a point singularity in the finite past, identified as Creation, are actually in this respect not at all inferior to the steady-state theory. For they, too, undertake to give us what may be regarded as the "physics" of creation, i.e., specifications as to initial motions, material composition and the like. The complaint that they locate the creation at a singular point in the past, thus making it inaccessible, is justified if it emphasizes that the creation *process* is left shrouded in mystery. Hoyle, therefore, is quite correct when he remarks that "it is against the spirit of scientific inquiry to regard observable effects as arising from 'causes unknown to science' and this is, in principle, what creation in the past implies." [4] *But exactly the same criticism applies to the continuous creation theory*. By spreading creation out in time and space, there is no reduction in the mystery, since multiplication of the occasions of creation as contrasted with the single unique event leaves it open to exactly the same objections as the latter.[5]

Hoyle, in particular, would argue on what he takes to be pragmatic grounds that all we need concern ourselves with is the ability of the theory to make successful predictions, its ability to work well. "We do not ask," he tells us, " 'Where does gravitation come from?', or if we do, science supplies no answer. Or again, we do not ask, 'Why do electric and magnetic forces occur in nature?' Instead we ask the question 'How does gravitation operate?' 'How do electric and magnetic forces operate?' Science does not seek to justify the existence of gravitation and electromagnetism. What science

[4] Hoyle, F., *Mon. Not. Roy. Astr. Soc.*, 108 (1948) 372.
[5] Cf. Dingle, H., *Scientific Adventure* (London, 1952) 166.

does say is, 'If gravitation exists then it works like this . . .' or 'If electricity exists then it works like this . . .' Exactly the same situation applies to the creation of matter. We cannot say why matter is created or where it comes from, but we can say 'If matter is created continuously then it is created in such and such a way.'"[6] It is true, of course, that scientific explanations do not undertake to give justifications in the above-intended sense, and in this sense it would be unwarranted to ask for the "purpose" of creation. It is also true that in speaking of "gravitation," "electricity" and "magnetism" all that is effectively involved for physics is bound up with the equations expressing these ideas, the rules according to which these are to be interpreted and their capacity to link known facts or predict fresh ones in a satisfactory way. If we would extend, however, these same ideas to the present theory, it would at best direct our estimation of its worth to an examination of what the equations contain, the various properties assigned to matter and the way we are enabled on the basis of these formulae to systematize our subject matter. It would, however, *not* justify the conclusion that matter has been "created in such and such a way."

3. Creation and the Use of Scientific Method

In saying that matter is created, the authors of this theory are explicitly clear about the fact that it is an *ex nihilo* process. Bondi writes: "It should be be clearly understood that the creation here discussed is the formation of matter not out of radiation but out of nothing."[7] Hoyle says: "From time to time people ask where the created material comes from. Well, it does not come from anywhere. Matter simply appears—it is created. At one time the various atoms composing the material do not exist and at a later time they do."[8] Finally, McCrea, in expounding the theory asserts: "There can be no causal treatment in a physical sense of true creation. This is almost a matter of definition. If the creation of matter is caused, as is conceivable, by existing physical conditions, then the true creation is of those conditions, and we should not try to give a physical theory of that creation. But we are here regarding the creation of matter as itself spon-

[6] Hoyle, F., *New York Times Magazine*, loc. cit.
[7] Bondi, H., *Cosmology*, 144.
[8] Hoyle, F., *The Nature of the Universe* (Oxford, 1950), 125.

taneous, i.e., as something "given" and not to be treated causally." [9]

Now to claim that matter "simply appears," that it is a "formation out of nothing," that its appearance is "not to be treated causally," is simply to deny the *possibility* of achieving any kind of scientific explanation of its appearance. This is, however, clearly a species of dogmatism, the irrevocable claim to an *ignorabimus* which is incompatible with the spirit and method of scientific inquiry. It is one thing to say that we don't yet understand how a process takes place, even though we might adduce reasonable grounds for affirming the existence of the process. It is an altogether different matter to say that we shall *never* understand its mechanism. The latter violates a primary rule of science, as Peirce expressed it, "not to block inquiry."

Suppose that indeed it were established in some way that matter does appear in the manner and with the various properties as claimed. We should then be able to speak of this as a "law of nature" in one sense of that phrase, namely as a regularity found to hold in fact. But it is precisely as a regularity that one would look for its explanation. And this criticism holds regardless of the particular philosophy of science one adopts. It will commonly be granted that one of the primary tasks of science is the discovery of laws. On one account, such laws are basically generalizations. Even where science advances to the level involving the establishment of an intricate deductive system or logical hierarchy of such laws, those which are in a given system the premisses of the system are still essentially statements of regularity.[10] Another view, such as the one adopted and used in this book, takes the primary technique of scientific explanation to consist in the application of laws or theories regarded as modes

---

[9] McCrea, W. H., *Endeavour, loc cit.*, 7.

[10] E.g., a recent statement of this view: "To explain a law . . . is to incorporate it in an established deductive system in which it is deducible from higher level laws. To explain these higher-level laws is to incorporate them and the deductive system in which they serve as premisses, in an established deductive system which is more comprehensive and in which these laws appear as conclusions. To explain the still-higher level laws serving as the premisses in this more comprehensive deductive system will require their deduction from laws at a still higher level in a still more comprehensive system. At each stage of explanation a 'Why?' question can significantly be asked of the explanatory hypotheses; there is no ultimate end to the hierarchy of scientific explanation, and thus no completely final explanation." Braithwaite, R. B., *Scientific Explanation* (Cambridge Univ. Press, 1953), 347.

of representation of phenomena and involving characteristic techniques of calculation or rules of inference, while not themselves being facts or generalizations of facts.[11] On either view, however, science always leaves the way open for more adequate explanations, and does not regard any given explanation as final. If we proceed by looking for more and more inclusive generalizations, then no given regularity (generalization) is ultimate, i.e., in principle unexplainable; the demand always exists for finding a more inclusive generalization under which a given one may be subsumed, i.e., deduced and so explained. And if we consider the primary technique of explanation to reside in the use of laws or theories interpreted as essentially techniques of inference, then, confronted with a hitherto unexplained regularity, it will be the objective of inquiry to find a theory whose way of looking at phenomena and whose mode of operation will enable the regularity to be understood. (Here, the question as to what "explains" the theory is a meaningless one, since theories are not like generalizations and are not explained by subsumption under "wider" theories. At best the rules which characterize one theory are incorporated, i.e., added to other rules which yield a more refined theory.) Given, then, the statement of the regularity with which matter appears, there is no reason, no methodological warrant, for insisting that this is itself an ultimate fact, capable of explaining other facts but incapable of being explained itself.

A confirmation by way of illustration of the above general criticism is to be found in the recent work of McCrea and McVittie.[12] Both seek, on the basis of the main ideas of the steady-state theory, some way in which the creation process can be made genuinely intelligible—and thus not a creation process at all. The basic suggestion put forward by McCrea, and worked out in a slightly different manner by McVittie, is that on the basis of relativity as distinguished from Newtonian theory, one can look for a connection between stress and the "creation" of matter. With a "negative stress" as made possible in relativity theory, the creation of matter becomes the mass-equivalent of the work done by this negative stress in the expansion of the universe. With the possibility as allowed

[11] Cf. Toulmin, S., The Philosophy of Science, 42, 84 ff.; Watson, W. H., On Understanding Physics, 52 ff.
[12] McCrea, W. H., Proc. Roy. Soc. (A) 206 (1951) 562; McVittie, G. C., Proc. Roy. Soc. (A) 211 (1952) 295.

under relativity views of the conversion of stress into mass and vice-versa, the creation of matter is no longer an *ex nihilo process*. From a methodological point of view (whatever might be the technical value of these suggestions) one sees here at work the scientific mind typically engaged in finding connections rather than showing its willingness to accept some fact as inexplicable.

4. The Meaningfulness of "Creation"

It may be said, however, in rejoinder, that to refer to the possibility of subsumption under "wider laws" or "incorporation within a wider theoretical framework" is precisely to miss the whole point in speaking of creation at all, since whatever may be the case in other situations, here one intends by the very employment of this concept to underline the point that we are confronted with an ultimate fact, itself capable of serving as a fundamental premiss but, by the very fact of its being ultimate, incapable of being explained in any way, by inclusion within any wider or more basic framework. That this is an error has just been argued on the ground that there is no warrant for taking scientifically any statement of a regularity as absolutely basic and inexplicable. Further, as I shall now argue, the very use of the term "creation" in such an allegedly ultimate statement is vacuous. This has the result that it leaves the supposedly fundamental premiss itself crucially indeterminate. As we have seen, there is an essential difference between the following two statements: (1) matter in an elemental form and with the various other properties as specified by the theory *is found* in the universe, and (2) matter in an elemental form and with the various other properties as specified by the theory *is created* in the universe. The authors of the theory insist on taking the second expression as the correct one, but whatever strength there is in the content of their proposals and methodologic soundness in their procedure lie in actually using the first expression. To say that matter is found in the universe leaves open the possibility of explaining its appearance, whereas to say it is created not only denies such a possibility but also *employs a term without any significant content*. "I have often been asked," Hoyle tells us, " 'Where does the newly created matter come from?' This is also a meaningless question. [Like the question about the origin of the universe as a whole, according to him.] It is only because in everyday life people have got used to the idea that matter must be conserved. When a conjuror pulls a rabbit out of his hat we know that the rabbit did not suddenly come into existence at the moment we see it and therefore it makes sense to ask 'Where did the rabbit come from?' But if the rabbit were indeed created by the conjuror, it might make no sense at all to ask this question." [13] To which we must reply that, on the contrary, whether presented with rabbits or particles of matter it *does* make sense to ask 'Where did they come from?' To be told that where these are created it makes no sense to ask this question, is *to beg the question,* to assume the meaningfulness and legitimacy of referring to the process as one of *creation.* Now as a matter of historical and etymological fact "creation," of course, *does* have a meaningfulness and legitimacy of employment in *certain contexts.* But it has notoriously undergone a series of transmutations and corruptions of meaning of which indeed the latest instance of degeneracy is to be found in the commentary accompanying the scientific theory we are examining. What is particularly noteworthy of this employment is the fact that the *last vestige* of meaning borrowed from its primary context of usage has been removed. This primary meaning of the term is to be found in the domain of human craftsmanship where it refers to the process of making some article of use such as a watch or a chair. What "creation" refers to here is at once a familiar and accessible fact of experience. Individual human beings, by virtue of some relatively distinctive use of skill and imagination are able to manipulate and transform already existent materials to yield a product whose structure and function can be appreciated by themselves or other members of the human community. When so located in this primary context of usage, the term "creation" allows of a literal analysis into such components as an agent or creator, materials used, methods of transformation or "making," and the finished product with its identifiable design or use. This literal meaning of the term has undergone, however, a variety of analogical extrapolations or truncations, which, while occasionally allowing innocent metaphors, have, instead of bringing fresh insight into another area, helped to breed confusion and support pseudo-explanatory devices. Thus what is in its primary usage a concept referring to a familiar fact of human experience, becomes under the pressure of uncontrolled analogical thought either

[13] Hoyle, F., *New York Times Magazine, loc. cit.*

a myth made to support ambitious metaphysical schemes, or, as it is transformed still further by theology, a cardinal mystery. The extension of the concept of creation beyond the domain of human craftsmanship to serve as a basis for cosmological speculation is the source of perhaps the most influential and persistent tradition of metaphysical thought. Its classic source is to be found in Plato. It is from Plato that metaphysical idealism and supernaturalism derive their inspiration in constructing a cosmology. Plato himself makes a clear and explicit appeal to human craftsmanship as the "root metaphor" employed in constructing his own theory of the universe. Plato's use of the imagery of human art in the myth of the *Timaeus* is guided by the profound conviction that such intelligibility as the world possesses is, at bottom, a purposive one, an adumbration of the Ideal Good. The cosmogony he pictures is conveyed in terms of a story not intended to be taken literally in its details. The Demiurge as "creator," the "pre-existent, recalcitrant materials" (the realm of Necessity and Chance) transformed by "rational persuasion" and purposive craftsmanship modelled on the Ideal and issuing in a world which exhibits a "mixture" of reason and necessity—all of these are not to be found in the "creation" of the cosmos in any literal way as in the case with ordinary craftsmanship. Conscious and deliberate myth here serves a philosophic conviction (itself the projection of a blind faith) in the designful character of the universe, the negation of all that the materialists and Sophists of Plato's day had proclaimed. Theology simply carried forward what Plato had here begun. What had been a conscious myth now became a literally intended mystery. The Creator not only cannot be located in the familiar world, He is no longer merely a symbol. His existence, literally claimed, is a "transcendent" one and basically an article of faith. Similarly, the process of creation becomes a divine mystery, the most real of facts but shielded from human understanding.[14] Even Human creation becomes, through an inverting rationalization, no longer the primary and guiding image. Instead it is now *man's imitation* of the infinite divine capacity and as such accorded a secondary, derivative status.[15] Throughout there is the pervading faith, continuous with Plato's, in the purposive character of the universe and all that it contains. As we turn to

[14] Cf. Augustine, *Confessions*, Book 11, §§ 4, 5.
[15] Cf. Maritain, J., *Art and Scholasticism* (New York, 1930) 123 ff.

the present day, the manner in which we find creation appealed to in the steady-state theory is one which, in effect, carries this progressive mystification to its last stage. For all of the sustaining motives or analogical threads of comparison with art are gone. Scientific cosmology, of course, now not only makes no claims about the designful character of the universe; it also stops short of making any reference to the Creator or the process of His making. It is not even claimed that these are mysteries whose existence is to be believed in even though not understood. All that it would retain is the fact that matter in an elemental form is created continuously. But if the Maker, the process of making, and the purpose are gone, *what is there left to the concept of creation?* Doesn't the very description of the appearance or presence of matter as one which is due to creation lose *all* its significance? Isn't it a case of having lost not merely its primary meaning, but even its various attenuated analogical modifications as in metaphysics and theology? If the *sole content* of the concept of creation is now simply that matter appears or is present, then far from this being a case of creation, it is at best, as previously suggested, a fact which invites scientific explanation.

5. The Perfect Cosmological Principle

We turn, finally to an argument of a methodological kind offered by Bondi in justification of the use of the concept of creation. Here it is important to note that the types of argument employed on the one side by Hoyle and on the other by Bondi reflect two radically different philosophies at work, reflected in the way each proceeds to build up the theory, though the end results are essentially similar. Hoyle, who works out his proposals within the framework of the field equations of the general theory of relativity, introduces a modification in the expression of those equations which allows him the opportunity of developing a cosmological model different in crucial respects from those hitherto encompassed within the gamut of "relativistic cosmologies." The primary justification in his eyes for entertaining both the modifications in the equations and the resultant cosmology is the fact that it can be submitted to the test of prediction. We have already seen that this pragmatic appeal, correct and important as it is, in no way warrants the interpretation, if the theory proves successful, that what has been confirmed is the claim that matter has been *created*. Meanwhile Bondi develops the theory along lines strongly reminiscent in some respects of the *"a priori* deductive" approach of Milne.[16] He would develop the cosmology not with the aid of the relativistic field equations but on the basis of what is regarded as a crucial principle, the *Perfect Cosmological Principle*. It is this principle which, it is claimed, warrants the introduction of the concept of creation.

The Perfect Cosmological Principle may be formulated briefly as the statement that "apart from local irregularities the universe presents the same aspect from any place at any time." [17] Restriction to the case of spatial homogeneity, as is usually the case with the great variety of current cosmological models other than the steady-state one, gives what is called the "narrow cosmological principle." According to it, all positions, *in space* are regarded as equivalent, from any of which a description of the universe can be made, since such descriptions will agree with one another.[18] Bondi would widen the principle to include a homogeneity or equivalence of times as well. The warrant for adopting the widened or "perfect" principle is twofold, according to him. In the first place, unless the principle were adopted, there would be no justification for assuming the general validity of physical laws. Taking laws as generalizations, the argument maintains that all ordinary physical science rests upon the basic axiom of the "unrestricted repeatability of all experiments." The repetition of an experiment and the expectation that it will yield the same results as the law specifies, assumes that change of place and time in the performance of the experiment will have no effect upon the result. "We see, therefore, that in all our physics we have presupposed a certain uniformity of space and time; we have assumed that we live in a world that is homogeneous at least as far as the laws of nature are concerned. Hence the underlying axiom of our physics makes certain demands on the structure of the universe; it requires

---

[16] Cf. chap. VI above.
[17] Bondi, *Cosmology*, 12.
[18] More accurately: "All large-scale averages of quantities derived from astronomical observations (i.e. determinations of the mean density of space, average size of galaxies, ratio of condensed to uncondensed matter, etc.) would tend statistically to a similar value independent of the positions of the observer, as the range of the observation is increased; provided only that the observations from different places are carried out at equivalent times." Bondi, *Mon. Not. Roy. Astr. Soc.* 108 (1948) 253.

a cosmological uniformity." [19] Secondly, any attempt to apply the generalizations won on the basis of terrestrial experience to vaster regions of space and time and ultimately to the structure of the universe as a whole, requires some justification for extrapolating such generalizations. Here reliance is made on an argument originally due to Mach, and illustrated in connection with the dynamical fact of rotation, that there is a strong "coupling" between the outcome of terrestrial experiments and the distribution of distant matter, e.g., in the system of the fixed stars. It is maintained in a general way, similarly, that "we can hence not contemplate a laboratory which is shielded to exclude all influence from the outside; and for the same reason we cannot have any logical basis for choosing physical laws and constants and assigning to them an existence independent of the structure of the universe." [20] Only some general assumption about the character of the universe at large will permit the use of laws and constants as holding without qualification throughout all regions of space and time. Here, moreover, instead of assuming that the laws and constants themselves undergo change, due to a general "evolution" of the universe, which would require—in order to make such a change or evolution itself meaningful and specifiable—the arbitrary stipulation that *some* laws or constants are indeed *invariable,* the proposal is made to avoid all such arbitrariness by rejecting all reference to an evolving or changing universe. This is accomplished by the Perfect Cosmological Principle which postulates that the universe is homogeneous and stationary in its large-scale appearance as well as in its physical laws.[21]

Use now is made of the cosmological principle in conjunction with the observed fact of local thermodynamic disequilibrium (the fact that more energy is found to exist in the form of matter than in that of radiation, as well as the fact that more energy is being radiated than is being absorbed by matter) to yield the conception of an expanding universe, a conception also supported by the observed recession of the nebulae. However, in order to satisfy the perfect cosmological principle, which requires a stationary (but not static) universe, together with expansion, there must be an abandonment of the principle of hydrodynamic continuity. "By

[19] Bondi, *Cosmology,* 11-12.
[20] Bondi, *Mon. Not. Roy. Astr. Soc.,* 108 (1948) 253.
[21] *ibid.,* 254.

the perfect cosmological principle the average density of matter must not undergo a secular change. There is only one way in which a constant density can be compatible with a motion of expansion, and that is by the *continual creation of matter.* Only if the diminution of density due to the drift to infinity is counteracted by a constant replenishment of newly created matter can an expanding universe preserve an unchanging aspect." [22] The above argument involves two essential features of philosophic significance, one having to do with the logical status of the Perfect Cosmological Principle, the second with the "deduction" allegedly made from it to the fact of creation.

The account given of the Perfect Cosmological Principle is in many ways strongly reminiscent of traditional discussions of the doctrine of the Uniformity of Nature offered as a solution to the problem of induction. As such, it suffers from precisely the same general difficulties that are already well recognized to hold for that more familiar formula, among them the fact that as a generalization it has whatever weaknesses are claimed to belong to other generalizations. It cannot be known with certainty if its truth-value is in any way dependent upon an appeal to experience. If the demand is not for certainty, there is no need for the Principle since any ordinary generalization involves inevitably the hazard that it will not continue to hold in instances beyond those already examined. Its truth-value, now construed as its probability, is no way altered by the appeal to the essentially vague principle of the Uniformity of Nature. If experience, on the other hand, does not enter into its establishment, then one must claim for it self-evidence, but this notoriously fails as a criterion of truth. Interpreted in its most favorable light, the Perfect Cosmological Principle, like the Principle of the Uniformity of Nature, functions not as a factual statement at all, capable of serving as a premiss in an argument, but as a definition that functions as a criterion or rule of what in the language of science is to be regarded as a law. To be a law, the rule now asserts, is to be a statement which by its very *meaning* asserts a structural connection among a selected number of factors. This connection *could not be otherwise* than it is at various places or times without surrendering its own distinctive and individual nature. The law, consequently, *could not* change or be different; at best we should employ *another* law and this in turn asserts a specific invariance or uniformity. So regarded, the Perfect Cosmological Principle is, to be sure, essential to science. However, it is not to be regarded as providing for other sciences a logical underpinning which is fathered upon cosmology. Cosmology shares such a principle equally with other sciences, since it is but a formal principle, a rule of the game, a defining characteristic of the techniques of representation which science employs regardless of its subject matter.

To say, moreover, that the Perfect Cosmological Principle is a formal principle means that it must surrender all power to serve as a premiss *in* an argument or to yield results of a factual kind. Inasmuch as it is a rule governing the formulation and use of laws or theories, it cannot be regarded as *part* of a theory in cosmology. This, in effect, means that the criticisms made, for example, of other cosmological theories in its behalf (as supposedly the unique possession of the steady-state theory) are not warranted, and it also means that, in particular, the attempt to deduce from it, among other things, the existence of a continuous creation of matter is unjustified.

We must reject the claim that "only in such a universe [the steady-state] . . . is there any basis for the assumption that the laws of physics are constant, and without such an assumption our knowledge, derived virtually at one instant of time must be quite inadequate for an interpretation of the universe and the dependence of its laws on its structure, and hence inadequate for any extrapolation into the future or the past." [23] The steady-state theory is really in no better situation, methodologically speaking, than those theories which, for example, posit an "evolution" of the universe. For *any* theory, it is necessary to specify some relationships as invariant. If what are regarded as "constants" in one theory are regarded as "variables" in another, then in turn new constants must be set up to give the treatment some determinate form. Thus to be an item in an evolutionary process is to forfeit the status of being a law or constant. Only what expresses the structure of this process is entitled to this status. Now whether an evolutionary cosmology or a steady-state one is to be regarded as successful, cannot be settled by saying that for all theories but the steady-state one, the selection of laws is arbitrary. For in one sense *any* theory, by claiming certain rela-

[23] Bondi, *Mon. Not. Roy. Astr. Soc.*, 108 (1948) 254.

tionships to hold and not others that are logically possible, is arbitrarily selective. Such selection must be justified now in the usual way by estimating the fruitfulness of its explanations and predictions.

Finally, no factual consequences such as are claimed to follow about the creation of matter can be drawn from the Perfect Cosmological Principle. To begin with, whether the universe is in a state of thermodynamic disequilibrium, or whether it is undergoing expansion, is something which we may claim to be the case or not, depending upon whether we take certain *arbitrarily selected* laws as holding in the interpretation of given observational data. These laws, such as the Doppler principle, or the various laws of thermodynamics, or atomic physics, that formulate the relations between matter and radiation, energy, entropy and the like, are, at best, useful devices, but by no means unique, i.e., without logically possible alternatives. The Perfect Cosmological Principle is not joined with these laws as another premiss to yield the result that the universe as a whole is in a state of thermodynamic disequilibrium or expansion. One uses the laws themselves that state the properties of thermodynamic disequilibrium or expansion in the interpretation of the data, but there is no requirement that they *must* be used. Finally, even were such interpretations to prove fruitful, it does not follow that the interpretation of the universe as being in a steady-state requires the creation of matter as a necessary condition. It would be sufficient for the purpose of the steady-state theory to propose that the average density of matter be constant, without presuming to offer in *that* theory any explanation for the appearance of matter so invoked. To provide such an explanation might be left for another theory of "finer grain" that might be forthcoming, without in any way weakening or causing the abandonment of the steady-state theory. One would thereby eschew dogmatism and the surrender of the search for intelligibility that is involved in the appeal to "creation."

Chapter 10

Scepticism

1. The Challenge of Operationalism

SCEPTICISM CONCERNING the possibility of attaining knowledge of the universe as a whole that might satisfy man's persistent craving in this direction is itself an ancient and persistent tradition. There have been many forms in which such scepticism has been voiced. While some have stemmed from theologically inspired motives for discouraging all efforts by man to probe the mysteries of the universe, others have appealed to the inherent limitations in the capacities of the human mind and the insuperable difficulties in the subject matter as grounds for such scepticism.

One recent expression of such scepticism stems in fact from the appeal to what is regarded as the essence of scientific method, namely, the use of "operational" techniques in the domain of cosmology. Bridgman writes: "Practically every one of our physical concepts demands the performance of an experiment, which in the first place can be indefinitely repeated in time, and which in the second place involves dividing the universe into two parts, one isolated from the rest, on which experiments are made by an external agency, whose actions are supposed arbitrary and unaffected by what occurs inside the isolated region. This procedure evidently breaks down when the subject is the entire universe. How, for example, shall we define the mass of the entire universe to the satisfaction of a critic who insists that the mass of the whole is not the sum of the masses of the part, a fact which we ourselves very well recognize even in small parts of the universe when we make our measurements accurate enough." Bridgman concludes that "the essential limitations of the experimental material place cosmology in a class by itself. It partakes as much as possible of the nature of the completely experimental subjects physics and chemistry, but is compelled by necessity to introduce features relating to less sharply defined human activities verging into the artistic, the

149

emotional and the metaphysical." [1] To this type of scepticism it is sufficient to point out that the criticism itself rests upon a misconception of what the resources of the scientific method are. Such a philosophy when carried to extremes overlooks the way in which theory functions over and above "operational" criteria, not only in cosmology but in physics and indeed all other advanced forms of science. It is simply not true that "practically every one of our physical concepts demands the performance of an experiment" since experiments, where they are performed, bear not on individual concepts but on a whole system of propositions. What is at stake in cosmology, for example, is not an individual concept like "the universe as a whole" or "the mass of the universe" for which we seek some "experimental" correlates, but a total theory which employs such concepts, and for such a theory the empirical consequences can be, in principle, just as readily appealed to in order to test its predictions as is done in any other branch of physics.

2. Kantian Dialectic

Kant's views on the nature of cosmology as stated in the *Critique of Pure Reason* is a classic statement of what, regarded from one point of view, is another form of scepticism which might be supposed to offer still a serious challenge to any type of conception of cosmological inquiry that foresees the possibility of positive results being achieved. One of the central contentions of his analysis is the claim that the mind inevitably falls into hopeless antinomies whenever it undertakes to determine anything definite about the structure of the universe as a whole. In the first of the four antinomies that he considers, that which relates to the space and time features of the universe, Kant seeks to show that one may with equal plausibility uphold two contradictory positions: that the universe is finite in space and had a beginning in time (thesis) and that it could not be either finite in space or in time, but on the contrary must be infinite in both respects (antithesis). For Kant the underlying difficulty posed by the antinomy is to be avoided by adopting as a solution the view of "critical" philosophy which bids us abandon any hope of determining whether the series into which we cast our representations of the world is either finite or infinite.

---

[1] Bridgman, P. W., "Nature and Limitations of Cosmical Inquiries," *Reflections of a Physicist* (New York, 1950) 216.

The function of Reason should not be one of adopting any positive dogmatic view as to whether the universe is finite or infinite. It is rather the legitimate function of Reason to prescribe a rule for the indefinitely continued search in "the regressive synthesis" performed by the understanding for further and still further terms in the spatial and temporal series in which the phenomenal character of the world is represented. Instead of pretending through a fallacious, because transcendent, leap to identify constitutively the world as it is in itself as an absolutely unconditioned totality, all that Reason can legitimately do is set as a continuing problem for the understanding the search for further terms in the series itself. And our understanding must never assume that at any given stage in its synthesis it has reached an absolute limit.

How shall such views be judged in the light of the method, results, and prospects of a scientifically oriented inquiry in cosmology such as we have been dealing with? Kant's views may be looked at in two different lights according to the emphasis we wish to give to the several elements in his thought. Our judgment as to whether such views are a direct challenge to the program of scientific cosmology or, on the contrary, may be brought into continuity with it, hinges on whether we take Kant as stressing primarily a philosophy of scepticism or, on the other hand, a conception of the method of science.

For Kant as a metaphysician in his own right the distinction between the regulative and constitutive uses of Reason is correlated with that between the two orders of being, the phenomenal and the noumenal. The former provides the domain for scientific exploration but is condemned to be incomplete; the latter while cognitively inaccessible is nevertheless taken as ultimate reality. Though we cannot have any constitutive knowledge of this hidden world, Reason nevertheless enforces upon (scientific) understanding an awareness of its phenomenal and essentially limited character. For Kant as a sceptic the attempt to deal with the notion of totality in his sense of an "absolutely unconditioned totality" is scientifically unwarranted because it becomes equivalent to the attempt to deal with a completed synthesis. This, he maintains, involves an illegitimate extension of a concept of the understanding. An unconditioned totality or completed synthesis is incapable of being experienced in any way, while to come within the range of actual or possible experience is a *sine qua non* of understanding. Such unwarranted use of the notion of totality would be present in either of two possible forms: where the series has an absolutely first term, and where the series as a whole without any first term is taken as unconditioned. These positions are adopted, respectively, by those who assert that the world had a beginning in time and a limit in space and by those who affirm its infinity in both cases. In the first case, it would be impossible to experience an absolutely first term. "Such an experience would have to contain a limitation of appearance by nothing, or by the void, and in the continued regress we should have to be able to encounter this limitation in a perception, which is impossible." [2] On the other hand, the claim that the universe is infinite is equally unwarranted since it is impossible to complete in experience by means of a successive synthesis, an infinite series.

The way in which in scientific cosmology the above difficulties as well as the necessity for recourse to the particular form of Kantian solution can be avoided is through a reformulation of the meaning of totality. In place of a *concept* of totality as based on the form of a *series,* one finds the introduction of a *theory* that deals with the universe as a *physical system.* To have a theory in cosmology means to bring to bear upon the facts of observational astronomy a pattern of ideas contained in a fundamental set of equations. To treat the universe as a whole as a physical system involves basically the same methodological apparatus as in any other branch of physical science. The scale is vaster but the basic techniques are the same. The use of physical theory in cosmology brings to light an important point of contrast with the Kantian discussion. For Kant the test of cognitive adequacy of a concept is the extent to which it can be exhibited in what he calls intuition, that is, sense-experience. On the ground that one could not hope to have any sense-experience of the totality of the space-time series, it is regarded as transcendent and therefore metaphysical. [3] But this is to take a narrow and unduly cramped view of the relation between the interpretative and factual elements in scientific knowledge. It is not necessary, nor is it indeed even meaningful that one have a sense-experience of the universe as a whole,

[2] *Critique of Pure Reason,* B 545.
[3] *ibid.,* B 511-512.

provided some theory defining such a totality leads by way of deduction or prediction to the observation of particular facts that *do* lie within the range of our senses and instruments. No more than this type of relation to sense-experience is actually demanded of scientific theories in any domain. For Kant "totality" was never to be reached as a matter of principle. For cosmology as a science, "totality" is present as a concept in every one of its theories, since it is the very business of cosmology to undertake to give some account of the universe as a whole. It is not the case that totality is forever beyond its reach; on the contrary, it is always within the range of its formulations. To the extent that a particular cosmologic theory continues to function adequately, we are justified in regarding the structure which it articulates as constituting, because defining, an intelligible pattern of the universe as a whole.

The difficulties that confront cosmology belong to the order of technical limitations, complexity of factors to be taken into account, and relative paucity of observational data, rather than a complete inaccessibility or imperviousness to scientific treatment. Ordinary scientific caution enforces a recognition of the tentative and corrigible character of any results reached. In cosmology no more than in any other area, theory can not claim for itself perfect predictive adequacy, simplicity, or comprehensiveness. Since the conflict among scientific theories is in general not of an antinomial character, the differences among several models of the universe in cosmology are to be regarded as exhibiting typical and familiar grounds of conflict among competing accounts that are possessed of varying degrees of success with respect to comprehending a common subject matter. By contrast, the temper of the arguments presented by Kant in behalf of the opposing positions in the antinomy as well as that present in his own solution expresses a quality of supposedly permanent validity for the considerations offered. Actually, for all their avowed universal and necessary character, all such arguments reflect the use of evidence accepted at a given time either because of its supposedly common sense nature or because it rests upon some mathematical conception or scientific theory themselves thought to be unexceptionable. Kant's treatment of the problems of cosmology is thus dominated at once by his own adoption of the Newtonian conception of space and time as absolute and the use of such conceptions in the statement of the opposing views of the antinomy. In the case of an explicitly avowed scientific approach, however, there is a clear sensitivity to the demands of the relatively fluid character of the evidence, observational and mathematical, upon which theory depends for its support. For cosmology as a science this means at once that the whole level of discussion must be removed from the sphere of *a priori* metaphysics to the progressive field of research.

3. Conclusion

The sceptical position may be summed up in general by the statement: "We can never know whether the universe is finite or infinite in space and time." The position that takes its guidance from the procedures and results of scientific inquiry may be summed up thus: "We can never establish at any given time whether some account of the universe which happens to be preferred to other accounts proposed at the same time, will remain adequate in the face of continuing inquiry." It is most important to realize that these two positions are quite different and it is only a very crude interpretation which will cover both by the assertion that "we shall never know the truth." For a sceptical view assumes that there is some structure possessed by the real universe, only men cannot hope to know it. For the scientifically guided view, however, there is no independent thing to which we can refer as "the universe" and with which various accounts may be matched in order to determine their relative truth. There is only, on the one side, a growing mass of observational data and, on the other, the variety of theoretic schemes for its interpretation. The assumption that there is a universe studied by cosmology which must already have one or the other of the properties, finitude or infinity, with respect to its spatial and temporal extent, needs to be challenged as not in fact required by the operative procedures of scientific cosmology.

Index

Anaximander, 1, 7, 21-22
Anaximenes, 17
Archimedes, 53
Aristarchus of Samos, 24n., 26
Aristotle, 15, 16, 22-26, 54, 56, 67, 94
Atomists, 15, 22, 24, 29-30

Baade, Walter, 126
Babylonian myths, 18-20, **21**
Berkeley, George, 95
Bohr's theory, 133
Bok, B. J., 126n.
Bolyai, J., 110n.
Bondi, H., 35n., 36-37, 62, 64, 81, 126n., 134n., 135n., 137, 143-147
Boyle's law, 52-53
Braithwaite, R. B., 138n.
Bridgman, P. W., 149-150

Callipus of Cyzicus, 24, **25**
Campbell, N. R., 50
Carnap, Rudolf, 110n.
Cepheid variables, 35
Charles' law, 52-53
Charlier, C. V. I., 34-35
Clerke, Agnes, 32-33
Copernicus, N., 15, 25, 26-28
Cornford, F. M., 17, 18n.
Curtis, H. B., 32n.

Dalton's theory, 133
Democritus, 30
Descartes, René, 55-56, 81, 100
de Sitter, W., 36-37, 62, 74, 76, 129
Dewey, John, 55-56
Digges, Thomas, 27-28
Dingle, H., 136n.
Diogenes Laertius, 18n.
Doppler effect, 36, 77, 148
Dreyer, J. L. E., 24n., 31n.
Duhem, Pierre, 48, 77-78

Ecclesiastes, xiii.
Eddington, Arthur S., 36-37, 62, 79
Edwards, Paul, 8
Egyptian myths, 18
Einstein, Albert, viii, 33, 36-37, 39, 41, 42, 50, 59, 60, 61-62, 68, 74, 76, 93, 96, 97, 99-104, 127-128
Enuma Elish, 16, 19f.
Euclid, 104-105
Euclidean geometry, 34, 35, 101, 104-105, 106, 107, 109-111
Euclidean space, 33, 34, 98-99, 114-115, 128, 129
Eudoxus of Cnidus, 24, 25, 26

Faraday, M., 53
Friedmann, A., 74

Galilei, Galileo, 27, 28, 53, 56, 98
Gamow, G., 126n., 132-133
Gauss, C. F., 104, 105, 106
Genesis, Book of, 19
Gold, T., 36-37, 62, 134n.
Greek myths, 18-19
Grant, Robert, 32

Hastie, W., 30n.
Heath, T. L., 24n.
Heckmann, O., 37n., 75
Helmholtz, H. L. F. von, 105
Herschel, William, 30-32
Hertz, H., 96
Hesiod, 17, 19, 20, 22
Hipparchus, 26
Hofstadter, Albert, 8
Holmes, A., 74
Hook, Sidney, 7

Hoyle, Fred, 36-37, 62, 134n., 135n., 136, 137, 140-141, 143
Hubble, Edwin, 35-37, 62n., 73-74, 125-127, 130-131
Huggins, W., 32
Humason, Milton, 74, 125
Hume, David, 56

Ionians, 15, 17

Jacobsen, Thorkild, 19n.
Jaeger, W., 18n.
Jammer, Max, 93n.
Judaeo-Christian theology, 26
Johnson, F. R., 27n.

Kant, Immanuel, 29-30, 38, 56, 101, 102, 104n., 150-154
Kepler, J., 24n., 42

Lambert, J. H., 34
Leibnitz, Gottfried Wilhelm, 55-56, 83, 95
Lemaître, G., 36-37, 62, 136
Lobachevsky, N. I., 104-105, 106, 110n.
Lorentz, H. A., 96, 98, 124
Lubbock, C., 31n.

Mach, E., 96, 100, 145
Maritain, J., 142n.
Maupertuis, Pierre Louis Moreau de, 29, 30
Maxwell, J. C., 41, 59, 96
McCrea, W. H., 35, 134n., 137, 138n., 139
McVittie, G. C., 8, 35n., 36-37, 62, 130-131, 139
Menger, Karl, 110n.
Michelson-Morley experiment, 59, 97, 98
Milne, E. A., 35, 62, 70, 78-80, 81, 85, 86-91, 118, 136, 144
Minkowski, H., 98
Mosaic cosmogony, 25-26
Munitz, Lenore B., 8

Nagel, Ernest, 7, 8
Newton, Isaac, 15-16, 30, 33-34, 35, 41, 42, 44, 49, 50, 52, 53, 59, 60, 69, 71, 93, 95, 96, 97, 100, 139

non-Euclidean geometry, 99, 104-105, 106, 109-111, 114-115

Olbers, H. W. M., 34, 35
Omer, Jr., G. C., 72n.
Opik, E. J., 126n.
Orphic cosmogony, 15-16, 18

Peirce, Charles S., 138
Plato, 11, 22, 54, 94, 142-143
Poincaré, H., 101, 105, 106, 107, 110
Pre-Socratics, 15, 16-18, 22
Ptolemy, C., 25-27
Pythagoras, 18
Pythagoreans, 15, 22

Reichenbach, Hans, 105, 108, 111-112
Riemann, B., 104-105, 106, 114, 127-128
Robertson, H. P., 36-37, 72n., 74, 88-89, 105, 113n., 114-115, 130n.
Rosen, E., 27n.
Rosse, Earl of, 31-32

Schwarzschild, K., 44, 68-69, 105, 106
Schweikart, F. K., 105, 106
Seeliger, H., 33-35
Slipher, V. M., 74
Smart, W. M., 132
Spinoza, Benedict, 101, 102, 104n.
St. Augustine, 133, 142n.
Struve, F. G. W., 31n.

Thales, 17
Timaeus, 11, 22, 142
Tolman, R. C., xiii, 36-37, 126n., 130n.
Törnebohm, H., 60n.
Toulmin, S., 43, 139n.

Watson, W. H., 49n., 139n.
Weyl, H., 128-129
Whitrow, G. J., 37n.
Whittaker, E. T., 133
Wright, Thomas, 28-29